This book is to be returned on or bef:
the last date stamped below.

GENE
REARRANGEMENT

Frontiers in Molecular Biology

Series editors

B.D.Hames

Department of Biochemistry, University of Leeds, Leeds LS2 9JT, UK

D.M.Glover

Cancer Research Campaign, Eukaryotic Molecular Genetics
Research Group, Department of Biochemistry, Imperial College
of Science and Technology, London SW7 2AZ, UK

Other titles in the series:

Genes and Embryos

Molecular Immunology

Molecular Neurobiology

Oncogenes

Transcription and Splicing

GENE REARRANGEMENT

Edited by

B.D.Hames

Department of Biochemistry, University of Leeds, Leeds LS2 9JT, UK

and

D.M.Glover

Cancer Research Campaign, Eukaryotic Molecular Genetics
Research Group, Department of Biochemistry, Imperial College
of Science and Technology, London SW7 2AZ, UK

IRL PRESS
—at—
OXFORD UNIVERSITY PRESS
Oxford New York Tokyo

Oxford University Press
Walton Street, Oxford OX2 6DP

Oxford is a trade mark of Oxford University Press

Published in the United States
by Oxford University Press, New York

British Library Cataloguing in Publication Data
Gene rearrangement.
  1. Genes. Recombination
  I. Hames, B.D. (B. David) II. Glover, David M. III. Series
  574.87322
  ISBN 0-19-963050-X
  ISBN 0-19-963051-8 pbk

Library of Congress Cataloging-in-Publication Data
Gene rearrangement/edited by B.D. Hames and D.M. Glover.
(Frontiers in molecular biology)
  Includes bibliographical references.
  1. Genetic recombination. 2. Gene expression. 3. Gene
  amplification. I. Hames, B.D. II. Glover, David M. III. Series.
  QH443.G44 1989   574.87'322—dc20   89-20021
  ISBN 0-19-963050-X
  ISBN 0-19-963051-8 (pbk.)

Previously announced as:

ISBN 1 85221 141 5
ISBN 1 85221 142 3 (Pbk)

Typeset and printed by Information Press Ltd, Oxford, England.

Preface

DNA sequence rearrangements are widespread in both prokaryotes and eukaryotes and are the subject of intense current research effort. Indeed the topic scope is now so large that it is impossible to review in any depth in a single book the recent discoveries which have been made in the field as a whole. An earlier volume in this series was devoted solely to the reorganization and expression of genes of the immune system. The present book extends this interest into several other related areas. Lex Van der Ploeg's chapter describes the molecular basis of antigenic variation in trypanosomes, covering DNA recombinational mechanisms in differential VSG gene expression in detail as well as other important expression events including discontinuous transcription and *trans*-splicing. In both the trypanosome and the immune system, DNA rearrangements are geared to providing a wide range of qualitative structural changes in the expressed gene product. In other circumstances it is the relative amount of a gene product which is the limiting factor, where the need for a particular product cannot be met by the maximal expression of the existing DNA sequences. Here DNA amplification is a solution to the problem. The chapter by George Stark, Michelle Debatisse, Geoffrey Wahl, and David Glover considers both developmentally programmed DNA amplification, particularly in dipteran flies where most progress has been made, and other increasingly important cases of amplification which occur as more isolated events such as the amplification of drug resistance genes and oncogenes in mammalian cells and the amplification of genes for resistance to toxic agents in whole organisms. Finally, every practising molecular biologist is all too aware that deletions, amplifications, inversions, insertions of foreign DNA, and translocations also occur in bacteria. We feel sure therefore that the chapter by Doug Berg, which examines the diverse mechanisms whereby these DNA rearrangements occur in prokaryotes, will be fascinating to a large cross-section of readers and not only to the specialists in this field. We thank each of the authors for their hard work and enthusiasm throughout this project and hope that their efforts will be as well-received by the readers as they were by the editors.

David Hames
David Glover

v

Contributors

Douglas E.Berg
Department of Microbiology and Immunology and Department of Genetics, Box 8093, Washington Medical School, St Louis, MO 63110, USA.

Michelle Debatisse
Unité de Génétique Somatique, Institut Pasteur, 25 rue du Dr Roux, 75724 Paris Cedex 15, France.

David M.Glover
Department of Biochemistry, Imperial College of Science and Technology, London SW7 2AZ, UK.

George R.Stark
Imperial Cancer Research Fund, Lincoln's Inn Fields, London WC2A 3PX, UK.

Lex H.T.Van der Ploeg
Department of Genetics and Development, College of Physicians and Surgeons, Columbia University, 701 West 168th Street, New York, NY 10032, USA.

Geoffrey M.Wahl
The Salk Institute for Biological Studies, San Diego, CA 92138, USA.

Contents

Abbreviations

ACE	amplification control element
APRT	adenine phosphoribosyl transferase
BC	basic copy
BrdU	bromodeoxyuridine
DHFR	dihydrofolate reductase
DM	double minute
EBV	Epstein-Barr virus
ECR	extended chromosomal region
ELC	expression-linked copy
ES	expression site
ESAG	expression site-associated gene
IHF	integration host factor
Int	integrase
IS	insertion sequence
LDL	lipoprotein lipase
MDR	multidrug resistance
medRNA	mini-exon donor RNA
MTX	methotrexate
mVAT	metacyclic VAT
NO	nucleolus organizer
PALA	N-(phosphonacetyl)-L-aspartate
PARP	procyclic acidic repetitive protein
PFG	pulsed-field gel electrophoresis
REP	repetitive extragenic palindromic sequences
snRNA	small nuclear RNA
snRNP	small nuclear ribonuclear particles
TK	thymidine kinase
Tn	transposon
VSG	variable cell surface glycoprotein
VAT	variant antigen type

1

Genomic rearrangements in prokaryotes

Douglas E.Berg

1. Introduction

Deletions, duplications and further gene amplifications, inversions, insertions of foreign DNAs, and translocations of resident DNA sequences to new sites have all been well documented in prokaryotes. These various DNA sequence rearrangements arise by diverse mechanisms: classical homologous recombination between repeated sequences; site-specific recombination; the movement of transposable elements; and several 'illegitimate recombination' processes. Certain rearrangements are as rare as point mutations whereas others are much more frequent, often developmentally regulated and, at the extreme, induced in each cell in a population by specific environmental or physiological signals. DNA rearrangements are of great biological interest. Some result from errors in DNA replication or repair or the movement of transposable elements, and provide models for frequent types of germ line and somatic mutations in humans. Some speed bacterial evolution, for example by facilitating the flow of genes among unrelated species and by generating duplications in which DNA sequences can diverge without selection and acquire new functions. Some contribute to pathogenicity or are part of a developmental program of prokaryotic cell differentiation. Some chromosomal inversions that were anticipated have not been found; such apparently 'forbidden' inversions may give insights into forces that determine the overall organization of the bacterial genome. Finally, some foreign DNAs seem prone to rearrange when cloned in *Escherichia coli*, an outcome of general significance to modern, recombinant DNA-based molecular biology.

2. Homologous recombination and genome organization

Repeated DNA segments in the size range of 1–10 kb collectively constitute several per cent of *E.coli* chromosomal DNA, and include various insertion sequence (IS) transposable elements, ribosomal DNA genes, and fragments of defective or cryptic prophages. Each of these repeated DNAs is normally present at less than 10 copies per genome (1–3). Repeated DNAs that are matched for at least 20–30 bp constitute

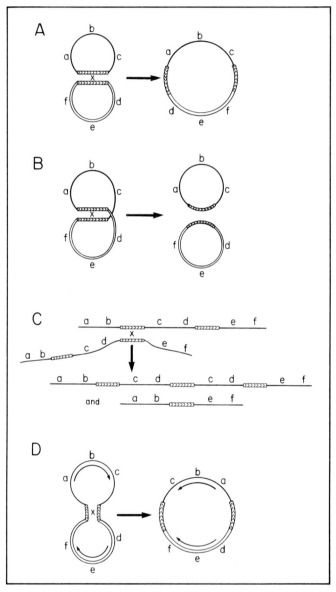

Figure 1. Consequences of recombination between homologous (repeated) sequences (━━). (A) Intermolecular recombination; (B) Intramolecular recombination involving direct repeats; (C) intermolecular recombination between direct repeats ('unequal' crossing-over); (D) intramolecular recombination involving inverted repeats.

substrates for classical homologous crossing-over mediated by the protein products of *recA* and of several other recombination genes (4,5). *Figure 1* summarizes how the positions and arrangement of homologous sequences determines the type of rearrangements formed by crossing-over between them.

2.1 Intermolecular recombination

A single reciprocal cross-over between two circular DNA molecules generates a single large molecule, a co-integrate containing both parental DNAs (*Figure 1A*). This is illustrated by the formation of Hfr strains of *E.coli*: the conjugative F factor plasmid present in F$^+$ strains contains copies of the insertion sequences IS2 and IS3; these elements are also present at multiple sites in the bacterial chromosome. Crossing-over between IS2 or IS3 in F and in the chromosome generates Hfr strains (co-integrates) at frequencies of about $10^{-3} - 10^{-2}$ (6,7). Many different Hfr strains have been found, differing in the site of F insertion and in the orientation of F in the chromosome, which reflects in part the many chromosomal positions of IS2 or IS3. (Additional Hfr strains arise by transposition, as detailed in Section 4.2.) Equivalent intermolecular recombination, but involving λ and pBR322-derived plasmid DNAs, has been used to select specific clones from recombinant DNA libraries as an alternative to screening methods such as plaque hybridization (8), and also to estimate the length of homology needed for recombination (9).

Recombination can affect the arrangement of sequences in multicopy plasmids such as pBR322 (10) and λ*dv* (an autonomously replicating deletion derivative of phage λ; 11, 12). Crossing-over between monomers as in *Figure 1A* generates dimeric forms of the same plasmid; higher oligomeric forms can arise by further recombinational interactions (13,14). The number of plasmid replication origins per cell tends to remain constant, so that plasmid oligomerization decreases the number of separate plasmid DNA molecules and when not checked can foster the segregation of plasmid-free cells (13).

The first λ*dv* plasmid was isolated in the late 1960s, before the ability of plasmids to exist stably in a dimeric form was recognized. Electron microscope measurements had indicated that the contour length of λ*dv* DNA was longer than expected from its known λ gene content. A reasonable interpretation was that some foreign DNA segments had been acquired during λ*dv* formation (11). Subsequent tests showed, however, that the predominant form in the *recA*$^+$ strain was a dimer, and that monomers generated by homologous recombination could be stably maintained in *recA*$^-$ cells (13,15). A selection that permitted new λ*dv* plasmids to be isolated easily was developed and used to study λ*dv* structure. In each case the new plasmids were found to lack foreign sequences (12,16).

Some pBR322-derived recombinant plasmids are found preferentially as monomers, whereas others are found preferentially as dimers in *recA*$^+$

cells. The predominant form is affected by the size of the monomer unit. In general, monomeric plasmids predominate when the size is less than about 5 kb, whereas dimers predominate when the size is in the range of 8 – 15 kb (17). The size distribution may reflect selection because the expression of certain genes is affected by the total size of the vector plasmid (C.M.Berg, personal communication).

2.2 Recombination between direct repeats

A single reciprocal cross-over between direct repeats within one DNA molecule generates two complementary smaller circular DNAs, each with a copy of the repeated sequence (*Figure 1B*). This is the major cause of reversion of Hfr strains of *E.coli* to F⁺ strains. Crossing-over involving other repeated DNAs can generate autonomous F′ plasmids in which chromosomal DNAs that had been adjacent to the inserted F remain linked to the autonomous F factor. The chromosomes in these F′ strains carry a complementary deletion (18,19). Equivalent deletions elsewhere in the chromosome must also arise by recombination, but they will not be detected if the product lacks a replication origin but contains genes needed for bacterial viability.

2.3 Unequal intermolecular crossing-over

A single cross-over between different copies of a directly repeated segment on different DNA molecules, or as drawn here (*Figure 1C*), in the two arms of a replication fork (unequal crossing-over), can lead to chromosomes with complementary duplications and deletions of the interstitial segment (the immediate product, upon completion of replication, is a dimeric molecule: a second cross-over can regenerate twin monomeric DNAs, one with the tandem duplication, and the other with the deletion). Further unequal crossing-over coupled with selection (e.g. for enhanced expression of a gene within the duplication) can lead to extensive amplification of the segment. Such amplified DNAs are inherently unstable; their extended homology makes them much more prone to become haploid by recombination than to arise in the first place (20).

2.4 Recombination between inverted repeats: 'permitted' and 'forbidden' chromosomal inversions

A single cross-over between inverted repeats reverses the orientation of the interstitial segment, but does not change the size or content of the complete DNA molecule (*Figure 1D*). Small inversions in the *E.coli* chromosome were easily detected in early studies by screening for derivatives of an Hfr strain which transferred chromosomal markers near its tail-end with extraordinarily high frequency (18). In contrast, a specific inversion of about one-fifth of the chromosome was not found after selecting recombination between a pair of mutant *lac* operons that were inverted and in different chromosomal locations (21,22). Inversions were also rare among illegitimate recombinants obtained by selection for fusion of a mutant promoterless *his* operon to other transcription units (23).

Recent systematic searches for inversions as products of homologous recombination have reinforced the view that many of them are in some way 'forbidden' (24–26). In these experiments transposons were used to place different mutant alleles of *lac*, *his*, or *tet* genes in opposite orientations at different chromosomal locations. Any Lac$^+$ (or His$^+$ or Tet^R) recombinant formed by a single reciprocal cross-over would carry a large inversion. The desired inversion derivatives were distinguished from non-inversion strains (that may have resulted from gene conversion) by testing for altered linkage of markers spanning the cross-over site (e.g. c and d in *Figure 1D*) (25), or by testing for an altered pattern of Hfr marker transfer in cross-streaking plate tests (24,26).

The distribution of 'permitted' and 'forbidden' inversions is generally correlated with the pattern of chromosome replication, which proceeds bidirectionally from an origin site to a termination region halfway around the chromosome. Most anticipated inversions with one endpoint in each replication arm (inversions of the origin relative to the terminus) were recovered, whereas many inversions entirely within one replication arm were not (25,26). Some inversions isolated on minimal medium proved to be lethal on rich medium. Others were so deleterious that they could be detected only in transient mating assays by an inverted pattern of Hfr transfer from young cultures, and were overgrown by fast-growing derivatives which had undergone an additional inversion and regained the standard gene arrangement (26).

A variety of models have been invoked to explain why certain inversions seem to be 'forbidden'. Some of them will be mentioned here.

(i)   Fitness or viability might depend on the relative dosage of certain key genes (25). Chromosomes are replicated continuously in rapidly growing cells, often with reinitiation of a new replication cycle before the previous cycle is complete, so that during rapid growth origin-linked genes tend to be 2- to 4-fold more abundant than terminus-linked genes. The lethality of inversions within one replication arm could reflect a deleterious effect of changes in the dosage of certain key genes or gene products.

(ii)   Fitness or viability might depend on the relative time of expression of certain key genes in the cell cycle. Transcription promoters that contain GATC (Dam methylation) sites are generally more active when hemi-methylated (just after passage of the replication fork) than when fully methylated (27,28). The lethality of certain inversions might thus reflect a change in the relative timing of expression of specific Dam-regulated genes.

(iii) The bacterial chromosome might be folded in a highly ordered structure by the binding of specific chromosomal loci on some sort of scaffold in a defined order. Inversions that disrupted the arrangement of chromosomal loci on this scaffold might be lethal (26).

(iv) Sequences at certain sites might be 'excluded' from the kind of single reciprocal cross-over that efficiently generates inversions by a mechanism that reflects the structure or replication pattern of the folded chromosome (25).

(v) Finally, the failure to recover some inversions and the rich-medium sensitivity and impairment of growth caused by others might reflect a transient cessation of replication each time RNA polymerase collides with the DNA replication machinery (29). In support of this model, the orientation of most very frequently transcribed genes matches that of replication fork movement whereas the orientation of most other genes is random with respect to replication. For example, 98 of 106 genes that specify components needed for protein synthesis are aligned with the direction of replication. The eight exceptions are small isolated genes which, because of their size, are probably often free of RNA polymerase (29). Many of the inversions that were sought but not recovered are also in accord with this model, although some inversions that apparently have been recovered (26) ought to have been lethal according to the polymerase collision model. At this point it seems that these apparent exceptions might reflect additional undetected rearrangements that compensated for the initial inversion.

2.5 Homologous recombination and phase variation: *Neisseria gonorrhoeae* pilin genes

Low levels of heterogeneity for certain surface structures in single clones are widespread in prokaryotes. Such heterogeneity may help them cope with variable and often hostile environments, for example immune responses in the animal or human host, or invasion of a population with virulent phage. The pili of *N.gonorrhoeae* provide the best known case of phase variation due to homologous recombination. Pili help bacteria attach to host mucosal surfaces; they are highly antigenic and in *Neisseria* are extremely variable in protein structure (30). Variants with new antigenically distinct pilins generally constitute 0.1 – 1% of the cells in young cultures grown *in vitro* without immune selection, and often predominate among re-isolates from infected humans.

Detailed analyses at the DNA level have shown that pilin variability is due to a recombinational scrambling of DNA sequences contributed by members of a large and divergent '*pil*' (pilin) gene family. Typical strains contain one or two copies of an expressed *pil* gene and multiple copies of divergent, but silent, and often truncated *pil* genes. Variant subclones arise by replacement of part of an expressed *pil* gene with the corresponding segment from a silent and partially homologous *pil* gene (30; *Figure 2*).

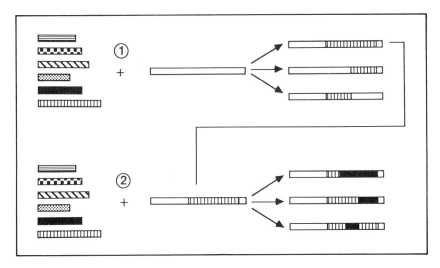

Figure 2. Pilin phase variation in *Neisseria gonorrhoeae*. Serial recombination events in regions of homology between divergent pilin genes (indicated by open and variously filled boxes) at silent loci and at the expressed locus generate a great variety of chimeric expressed pilin genes. Two successive recombination events, 1 and 2, are shown. (Reprinted from ref. 30, with permission.)

Initial studies suggested that allele replacement was not due to reciprocal recombination because in the variant strains the apparent donor locus did not contain sequences removed from the expressed locus; rather, sequences from the expressed locus were simply lost from the cell lineage. This seemed compatible with models of site-specific or transpositional recombination (31). *Neisseria* tend to lyse during growth and are also highly transformable, and it has been concluded that most *pil* gene variability is due to classical recombination between transforming DNA and endogenous chromosomal sequences, rather than between two loci in the same intact chromosome (32,33).

(i)   Switching of pilin types is *recA*-dependent, unlike most site-specific recombination.

(ii)   Reconstruction experiments using pilin genes marked with inserts of an antibiotic resistance gene showed that silent (incomplete) as well as expressed loci can undergo replacement.

(iii)   This variation is blocked by use of mutants that, although recombination proficient, are defective in DNA uptake during transformation, or by the addition of DNase to donor cells prior to mixing them with recipients (30,32,33).

3. Site-specific recombination

Examples of site-specific recombination reactions in prokaryotes include phage λ integration and excision, resolution of transpositional co-integrates, circularization of linear DNAs, monomerization of plasmid dimers, DNA inversion causing phase variation, and rearrangements implicated in cell differentiation. Each of these events entails protein-mediated DNA breakage at precisely defined sites and rejoining of broken ends without DNA synthesis.

3.1 Integration and excision of phage λ

The integration of λ into the *E.coli* chromosome and λ prophage excision provide well understood examples of how site-specific recombination can be sensitively regulated and, in turn, can affect key steps in a developmental pathway.

3.1.1 General features and an overview of λ

Integration and excision (*Figure 3*) constitute important stages in both lysogenic and lytic λ development (for reviews see refs 34 – 36). Integration results from a single reciprocal cross-over between a 240 bp 'attachment' (*att*) site in the phage (designated PP′ in *Figure 3*) and a 25 bp *att* site in the bacterium (BB′ in *Figure 3*). It is mediated by the λ-encoded integrase protein (Int) plus 'integration host factor' (IHF), which is composed of two host proteins. Int and IHF bind cooperatively at several places within PP′ and Int also binds BB′. Int protein makes a 7 bp staggered cut in a 15 bp segment that is identical in both *att* sites, twists the DNAs, and rejoins free 3′ and 5′ ends, thereby achieving integration. λ prophage excision also involves a single reciprocal exchange between the hybrid BP′ and PB′ sites. This is mediated by Int and IHF plus the λ-encoded excisionase (Xis) protein. At low Xis concentrations excision is facilitated by the host-encoded Fis protein. The complex containing Xis, as well as Int and IHF, is specific for the hybrid BP′ and PB′ sites formed by integration; it cannot recombine the PP′ and BB′ sites formed by excision.

There are many different phages related to λ including 21, φ80 and 434 of *E.coli* and P22 of *Salmonella typhimurium* (37 – 39). They resemble λ in the relative locations of genes and sites for functions such as integration, prophage immunity, DNA replication, and DNA packaging, but many of the corresponding segments differ in sequence and in the specificities of the proteins they encode. The lambdoid phages are distinct from P1 (which is normally maintained in lysogens as a non-integrated plasmid; 40), the mutator phage Mu (41; Section 4), and P2 (42), which belongs to another family of temperate phages.

The various lambdoid phages were initially distinguished from one another by immunity. A specific lambdoid phage will not grow on or lysogenize cells that already carry a prophage of the same immunity type (homo-immune), but each can grow on or lysogenize cells that carry a prophage of another immunity specificity (hetero-immune) (43), and can recombine with phage λ in regions of homology.

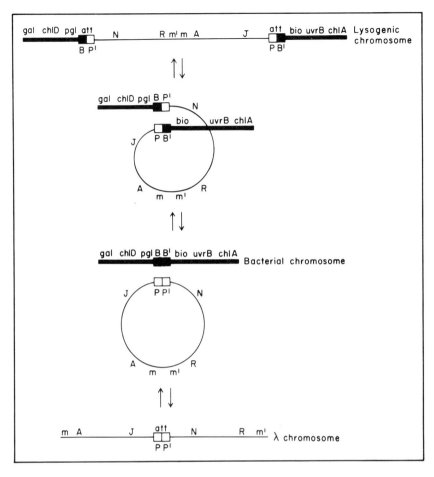

Figure 3.   Integration and excision of phage λ. A, J, N, and R are phage genes. PP′ and BB′ are the phage and bacterial attachment sites. m and m′ (often designated *cos* sites) are the ends of the λ DNA in virus particles. *gal*, *chlD*, *pglB*, *bio*, *uvrB*, and *chlA* are bacterial genes. λ DNA as found in phage particles is shown at the bottom. (Reprinted from ref. 36, with permission.)

Integration reactions are also highly phage-specific. For example, the Int protein of λ does not act on ϕ80, nor, conversely, does Int of ϕ80 act on λ. Each lambdoid phage also integrates preferentially at a specific site: the *att* site for λ is between *gal* and *bio*; the *att* site for ϕ80 is one-tenth of the bacterial genome away, near *trp*; the *att* site in *S.typhimurium* for P22 is at a third locus, near *pro*. Phage 434 integrates at the site used by λ although these two phages are hetero-immune, which underscores both the homology between the *int* genes of λ and 434, and the separate determination of integration and immunity specificities (38,39).

Recombination among different lambdoid phages is quite common. Interspersed with dissimilar or divergent sequences are homologous DNA

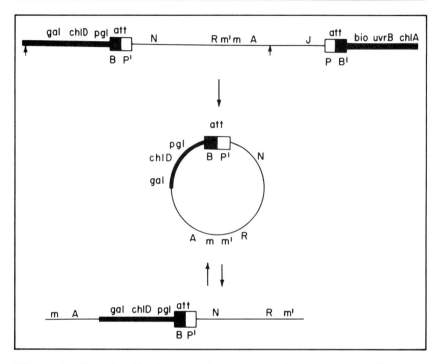

Figure 4.   Genesis of λ*gal* transducing phage by aberrant excision. Top line, integrated prophage. Upward arrows under top line indicate sites of aberrant excision. Bottom line, λ*gal*-transducing phage as found in virions. This transducing phage lacks the essential gene *J* and is thus defective for lytic growth unless complemented by a co-infecting helper phage. λ*bio* transducing phage would be formed by similar aberrant excision, but involving a site to the right of *bio* in bacterial DNA. See *Figure 3* legend for definition of nomenclature. (Reprinted from ref. 36, with permission.)

segments that can be hundreds or thousands of base pairs long and that constitute good recombination substrates. Genes and sites with related functions such as *int* and the PP' site, or the *cI* repressor gene and its operator sites are tightly linked. In consequence, crossing-over can generate new combinations of alleles (e.g. shuffling integration and immunity specificity), but tends not to disrupt co-adapted functional units (37,38,43). Recombinant phage best suited for a specific niche tend to be selected, a situation analogous to the recombinational scrambling of *Neisseria* pilus gene segments outlined in Section 2.5. The evolutionary advantage stemming from the exchange of functional units may help maintain a relatively uniform gene arrangement among phages. [The need for proper timing and extent of expression of individual genes during lytic growth (44) would also contribute to the maintenance of a single optimal genome organization.]

3.1.2 Polylysogens: multiple λ prophages in tandem

Lysogens carrying two or more complete prophages in tandem can arise in two ways.

(i)   Dimeric λ DNA can be integrated.

(ii) Int protein can mediate recombination between PP′ and BP′ or PB′ (prophage) sites (45). Hence tandem prophages can be formed by integration into a chromosome already carring one prophage (two separate integrations in one cycle of lysogenization, or integration after infection of a hetero-immune lysogen) (46).

Double lysogens are inherently unstable because crossing-over anywhere in the homologous sequences can result in loss of one prophage, or in amplification (*Figure 1B* and *D*).

3.1.3 Aberrant excision and specialized transducing phage

Specialized transducing phage carrying host genes that had been adjacent to one λ prophage end (e.g. *gal* or *bio*) arise by aberrant excision (*Figure 4*), and can be selected by transduction of appropriate (Gal⁻ or Bio⁻) bacterial host strains. Their formation does not depend on Int or Xis (36,45), and may result from DNA breakage and joining events, stimulated by replication of the induced prophage *in situ*. These phages carry bacterial DNA as substitutions adjacent to the hybrid phage attachment site, often replacing essential phage genes (*Figure 4*). In consequence, many transducing phage are defective, and unable to grow lytically unless complemented by a helper λ phage. They can lysogenize their bacterial hosts, however, by integration into the BB′ site (*Figure 5*), by integration into hybrid *att* sites in the case of lysogens (45), or by homologous recombination between shared bacterial or prophage sequences (*Figure 1A*). Any of these events duplicate chromosomal sequences, and thus increase the chance of further amplification or genome rearrangement.

3.1.4 Secondary site lysogens and the specificity of integration

E.coli strains lacking the primary *att* site can be lysogenized inefficiently by λ integration into other ('secondary') sites. These integration reactions also require Int protein and the PP′ site in λ. Hundreds of such secondary sites have been identified, and are used with efficiencies ranging from 10^{-8} to nearly 10^{-2} that of the primary site. About 1% of secondary site lysogens are auxotrophic or otherwise mutant due to the λ insertion (47,48).

The correct excision of λ from secondary sites is much less efficient than excision from the primary site, but can be scored by reversion of auxotrophic insertion mutations to prototrophy or by the loss of the λ prophage (e.g. using the temperature-sensitive growth phenotype of cells lysogenic for a thermo-inducible prophage). Excision regenerates the parental phage and bacterial sequences, and requires Int and Xis proteins. Aberrant excision from secondary sites generates phages carrying

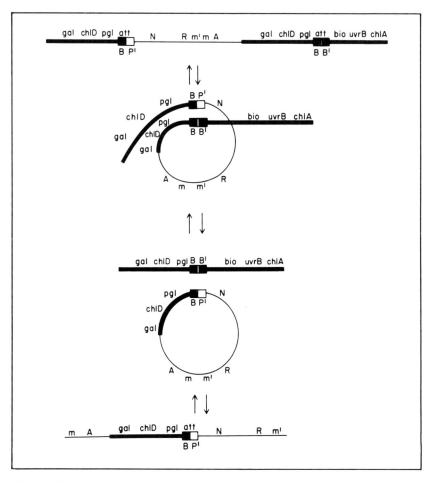

Figure 5. Integration and excision of λ*gal* phage. See *Figure 3* legend for definition of nomenclature. (Reprinted from ref. 36, with permission.)

adjacent bacterial DNAs that are equivalent to the classical λ*gal* (or λ*bio*) transducing phage shown in *Figure 4* (48,49). These new specialized transducing phage can also be incorporated into bacterial genomes by the action of Int protein, or by crossing-over in regions of phage or bacterial DNA sequence homology. The resulting duplications can foster further recombinational interactions and genome rearrangements. The specialized transducing phage formed by aberrant excision from primary and secondary sites are important historically, having permitted *in vivo* cloning of many *E.coli* genes in the years before *in vitro* recombinant DNA cloning became commonplace (48,49).

3.1.5 Regulation

The three choices for λ: (i) integration; (ii) persistence as an integrated prophage; and (iii) excision, are determined by the levels of Int, Xis, IHF,

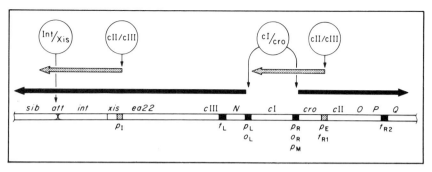

Figure 6. Regulation of transcription during the early stage of λ development. Immediately after infection, most transcription from p_L and p_R is terminated at t_L and t_{R1}. The N protein eliminates these termination events. The cII and cIII proteins channel λ development toward the lysogenic response because cII stimulates RNA synthesis from p_E and p_I (hatched arrows) and delays the expression of late lytic functions. Note that the transcript from p_I induced by cII/cIII terminates before (to the right of) the *sib* site. In contrast, transcription from p_L in non-integrated λ DNA is anti-terminated by N protein, and therefore continues through *sib*. The *sib* sequence is cleaved by RNase III, which permits exonucleolytic degradation of *int* mRNA sequences. (Reprinted from ref. 34, with permission.)

and Fis proteins. Sophisticated controls make these choices highly responsive to cell physiology and the environment, and coordinate the choices with separate commitments to lysogenic or lytic development. Site-specific recombination has long been appreciated for its role in permitting the virus to remain as a stable quiescent prophage. The recombination reactions also serve as biological switches that can help regulate the probability and timing of commitments to lysogenic or lytic growth, and help ensure that commitment to one of these fates, once made, is essentially irreversible (34,35).

Infection by two or more phages per cell favors integration and the establishment of repression (lysogenization) whereas single infection results in lytic growth. The λ cII and cIII proteins and the host Hfl protein comprise a regulatory circuit that makes the choice between these alternatives multiplicity-dependent. When cII protein accumulates to critical threshold levels it can induce Int and cI repressor synthesis by activating the promoters p_I and p_E, respectively (*Figure 6*), and thereby foster stable lysogenization. Competitive interactions involving Hfl and cIII proteins amplify the effect of phage multiplicity on cII protein levels: cII protein is inactivated by the host Hfl protein, which is in turn antagonized by cIII protein. cIII protein is also unstable, hence each component in this chain of interactions helps make the chance of lysogenization responsive to the phage population size.

Although the *int* gene is transcribed from the p_L promoter, several factors act cumulatively to prevent λ integration during lytic growth. (i)   Xis protein, which is also made from p_L (but not from p_I) inhibits the integration activity of Int protein (50)

(ii)   The *int* and *xis* genes overlap, and the translation of *xis* mRNA interferes with the translation of *int* mRNA

(iii) *int* mRNA is much less stable when made from the p_L promoter during lytic growth than when made from the p_I promoter. This instability is due to a sequence called *sib* that is transcribed from p_L, but not from p_I in non-integrated λ DNAs (see *Figure 6*).

The *sib* site and the *int* gene are on opposite sides of *att* (PP′), and hence are separated from each other by integration. Int as well as Xis is needed for prophage excision, and the prophage structure ensures that *sib* interferes with Int synthesis only after excision (34,35).

Severe damage to the DNA of lysogens by UV irradiation, chemical mutagens, and carcinogens results in a series of changes culminating in:

(i)   destruction of the prophage repressor
(ii)   transcription from lytic phage promoters including p_L
(iii) synthesis of Xis and Int proteins
(iv)   prophage excision, and lytic growth in every cell in the population.

Mild DNA damage or chance fluctuations in repressor levels can also lead to prophage induction, but with a probability that depends strongly on cell physiology. This effect is mediated by Fis, a host-encoded protein that facilitates excision when Xis is limiting (as in cases of partial derepression), but is not needed when Xis is in excess, and is abundant only when cells are growing rapidly. Thus, the conditional Fis dependence helps link the probability of sporadic excision to physiological conditions that are well suited for phage lytic growth (35).

3.2 Resolution of transposition co-integrates

Intermolecular transposition of transposon-*3* (Tn*3*)-related elements generates co-integrates in which donor and target DNAs are joined by direct repeats of the transposon (see Section 4). These co-integrates undergo site-specific recombination (termed resolution) that regenerates the donor DNA and also a free target DNA carrying a transposed copy of the element. Resolution is mediated by an element-encoded 'resolvase' protein which acts at a 120 bp sequence called *res* (51,52). Resolvase is related to a set of invertase proteins that show a complementary specificity for inverted repeats of their cognate recombination sites (see Section 3.4). The transposable elements IS*1* and Mu, although not in the Tn*3* family, also often generate co-integrates when they transpose. They do not specify equivalent resolvases, however, and the co-integrates they generate are broken down only by less efficient homologous recombination.

Resolvases exhibit remarkable specificity for *res* sites in the correct position and orientation. Studies with small plasmid substrates have shown that resolvases recombine direct but not inverted repeats of their cognate *res* sites, and that they do not act on *res* sites in separate DNA molecules (to regenerate co-integrates). Several models to explain this specificity have been considered. It now seems likely that the specificity may reflect the way in which *res* sites are bound by oligomers of resolvase protein, random collision between *res* site – resolvase complexes, and a high probability of productive interactions only when the *res* sites are linked as direct repeats in the same molecule (51,52).

The Tn*3* transposon family is large and diverse. Although the resolvases of these transposons are related to each other at the DNA sequence level, in most cases the resolvase encoded by one element does not operate on other members of the family (52). As a result of this specificity, DNAs can sustain inserts of several different members of the Tn*3* family without risk of resolvase-promoted deletions.

The Tn*3* resolvase (*tnpR*) gene promoter is within the *res* site. Resolvase therefore also serves as a transcriptional repressor, autoregulating its own synthesis, and thus controlling as well as mediating the resolution process. The Tn*3* transposase (*tnpA*) gene promoter is also within *res*, and hence resolvase regulates the synthesis of transposase and the frequency of Tn*3* transposition. $\gamma\delta$, an element closely related to Tn*3* that is present in F and implicated in many cases of Hfr formation in *E.coli* (see Section 4), is regulated similarly: the promoters for the resolvase and transposase genes are also within its *res* site (51,52). In contrast, resolvase does not seem to be autogenously regulated in members of other branches of the Tn*3* family. For example, the synthesis of resolvase and resolution of co-integrates of the mercury resistance transposon Tn*501* is induced by mercuric ion (which also induces transcription of the transposase and mercury resistance genes) (52).

3.3 Site-specific recombination and colE1 plasmid stability

The stable maintenance of colE1 and related multicopy plasmids depends on a site-specific recombination process that converts plasmid dimers formed by homologous recombination back to monomers (53). It occurs at a specific 240 bp site in colE1 called *cer*, and is mediated by the products of chromosomal, not plasmid-borne, genes. Three bacterial genes needed for recombination at *cer* sites have been found, called *xerA*, *xerB*, and *xerC*. The *xerA* and *xerB* genes have been characterized to some extent (54). *xerA* is identical to *argR*, which encodes the repressor of transcription of the arginine biosynthetic genes. The *xerB* product is related to a peptidase from *Salmonella*.

Fragmentary evidence suggests that the *xerA* and *xerB* proteins play a structural role in *cer*-recombination.

(i)   There is a XerA binding site in *cerA*, but it is 200 bp from the cross-over site (54)
(ii)   A mutation in *cer* that makes recombination independent of *xerA* and *xerB* was found to lie outside the XerA binding site
(iii) Finally, this XerA-independent mutation permits intermolecular

recombination (to generate dimers from monomers) (D.Summers, cited in ref. 54). It was postulated that XerA protein helps assemble *cer* sites into a synaptic complex whose structure ensures the observed directionality of recombination (54).

3.4 Cre-loxP recombination: P1 lysogenization and plasmid maintenance

Phage P1, unlike λ, is maintained in lysogens as a plasmid, not an integrated prophage, at about 1–2 P1 DNAs per chromosome (40). In *recA*[+] cells, that permit the formation of dimeric P1 DNAs, efficient partitioning of P1 genomes to both daughters at cell division depends on the action of the P1-encoded Cre protein, a recombinase that acts specifically at a site in P1 called *loxP* (55,56). Cre-mediated recombination is thus formally equivalent to the resolution of colE1 plasmid dimers and Tn*3* transposition co-integrates discussed above.

The *loxP* site is 34 bp long, and consists of 13 bp perfect inverted repeats bracketing an 8 bp asymmetric spacer in which Cre protein makes a 6 bp staggered cut. Mutational analyses have shown that the exact sequence of the inverted repeats is important for *loxP* function, whereas that of the spacer is not: *loxP* sites with variant spacers can recombine efficiently provided that the two spacers are identical in sequence. Cre-mediated recombination at *loxP* sites with symmetric spacers generates inversions and deletions with equal frequency, indicating that asymmetry in normal spacers, and thus probably complementary base-pairing of a recombination intermediate, determines the directionality of Cre-mediated recombination (40,57,58).

Cre-mediated recombination may contribute in additional ways to P1 fitness. P1 DNAs are packaged into virions sequentially from a concatemer at the end of lytic growth. The packaged DNAs are linear, terminally redundant, and circularly permuted. The DNAs in about one fourth of the virions contain direct repeats of the *loxP* site and are good substrates for Cre-mediated circularization. Cre[−] mutations cause a 10- to 25-fold decrease in efficiencies of lysogenization in a *recA*[−] *E.coli*, although they do not significantly affect lysogenizaton of a *recA*[+] strain (where the terminally redundant ends would be substrates for generalized recombination) (40,56,58). Cre protein may therefore facilitate lysogenization under conditions in which recombination is suboptimal.

Studies of P1-mediated integrative suppression of host *dnaA*[−] mutations indicated that *loxP* sites can also participate in other rearrangements. The DnaA protein is needed for the normal initiation of chromosome replication, and mutations in *dnaA* can be compensated by P1 insertion into the *E.coli* chromosome and initiation of replication from the P1 prophage origin. Sequence analyses have indicated that P1 insertion involves recombination between the *loxP* site in P1 and a *loxP*-like sequence in the bacterial chromosome (59,60).

3.5 DNA inversion and phase variation

Phase variation, as noted in Section 2.5, refers to a reversible switching between alternative phenotypic traits. It is found in a variety of organisms and is due to DNA inversion by site-specific recombination. DNA sequence comparisons and functional tests have identified two different families of invertase proteins, one related to the Tn*3* ($\gamma\delta$) resolvase, and a second to phage λ integrase (61).

3.5.1 Phase variation in flagellae of Salmonella

The protein composition of flagellae in lineages of *Salmonella typhimurium* are switched between type H1 and type H2 at frequencies of up to about 10^{-3} per cell per generation. This low level variation may contribute to a persistent infection by enabling some cells in each clonal population to elude host immune responses. Alternatively, because flagellae are used as receptors by certain virulent phages, phase variation may ensure that some members of any bacterial clone will survive epidemics caused by a flagellae-specific phage (61 – 63).

This variation results from the sporadic inversion of a DNA segment carrying a promoter for transcription of the genes for H2 flagellin and for a repressor of the unlinked type *H1* flagellin gene (*Figure 7a*). When the promoter is in one orientation, H2 flagellin and the repressor are made; when the promoter is inverted, synthesis of H2 and repressor cease and H1 flagellin synthesis begins. H1 flagellae continue to be made in the clonal descendants of this switched cell until by chance the promoter segment is once again inverted (61).

Inversion is mediated by a protein called Hin that acts on inverted repeats of a site called *hix*. Each *hix* site is 26 bp long and consists of imperfect 12 bp inverted repeats flanking a 2 bp spacer, which is the actual site of DNA breakage and joining. Hin protein is related at the sequence level to the resolvase of $\gamma\delta$, but it is specific for inverted repeats of its cognate *hix* sites: it will not recombine *hix* sites in different molecules nor direct repeats of *hix* sites in the same DNA molecule. Hin protein acts in concert with Fis protein [that also stimulates λ prophage excision (Section 3.1.5)]. Fis acts at a 60 bp 'enhancer' sequence that can stimulate inversion in either orientation and at essentially any distance from *hix* sites. The Fis protein – enhancer interaction is thought to help bring the *hix* sites into a nucleoprotein complex in which Hin protein can act.

3.5.2 Phase variation of the host range of phage Mu

Different individual Mu phage particles obtained by induction of a single lysogenic culture are able to infect either *E.coli* K-12 or *Citrobacter freundii*. These alternate host ranges reflect inversion of a DNA segment (called the G segment; *Figure 7b*) which determines the phage tail fiber type. The inversion site is within an open reading frame for S protein, one of the two proteins needed for Mu tail formation. The constant 5′ end of the

S gene is joined to the 3' end of S_v when G is in one orientation, and to the 3' end of S'_v, which differs in sequence, when G is in the reverse orientation. Genes *U* and *U'*, which are also needed for tail synthesis, are in opposite orientations within the G segment, and are expressed coordinately with *S* and *S'*, respectively (61,64).

G inversion is mediated by the protein product of the Mu *gin* gene which, remarkably, is closely related at the sequence level to the *hin* gene of

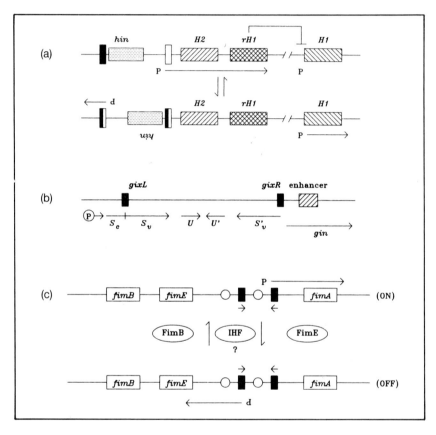

Figure 7. Recombination switches that underly phase variation phenomena. (a) Inversion-controlled expression of the flagellin genes of *S.typhimurium*. *H2* and *H1* encode flagellin proteins; *rH1* encodes a protein repressor of *H1* transcription; *hin* encodes the invertase protein (the sites on which it operates are indicated by the open and filled boxes). Top line, configuration leading to *H2* expression. Second line, configuration leading to *H1* expression. (b) Invertible region of phage Mu that determines phage host range. The G⁺ orientation, resulting in phage able to infect *E.coli* K-12 but not *Citrobacter*, is shown. Note that inversion changes the 3' end of gene *S* (fusing S_c to S_v or to S'_v) and also coordinately elicits expression of either gene *U* or gene *U'*, respectively. (c) Invertible on/off switch for type I pili. Pilin is encoded in gene *fimA*. This switch is under directional control: FimB protein is needed for off→on; FimE protein is needed for on→off. (Reprinted from ref. 61, with permission.)

Salmonella. Phage P1, which in most regions is unrelated to Mu, contains highly homologous tail fiber and invertase genes, and also exhibits host range phase variation. The Mu, P1, and *Salmonella* inversion systems are so closely related in sequence that a defect in one invertase gene can be complemented by wild-type alleles of either of the other two genes. The inversion of the Mu and P1 tail fiber gene segments is also stimulated by Fis protein acting at an enhancer equivalent to that used for Hin inversion. There is no evidence that the invertible segments of Mu, P1, or *Salmonella* are currently transposable elements, and the type of rearrangement events that incorporated them into the three different genomes are not known.

3.5.3 On/off phase variation for synthesis of pili by E.coli

The type I pili (fimbriae) of *E.coli* mediate adherence to host mucosal and phagocytic cells. Their synthesis is turned on and off by inversion of a DNA segment containing the pilin gene promoter at frequencies of about 10^{-3} per cell per generation. Pilin phase variation may contribute to the virulence of these organisms by facilitating the colonization of mucosal cells at the onset of infection, and the avoidance of phagocytosis by some descendant (*pil$^-$*) subclones in later invasive stages (61).

The switch from off to on is mediated by the product of the *fimB* gene; the reciprocal inversion from on to off is mediated by the product of *fimE* (65). Both the *fimB* and the *fimE* gene are related evolutionarily to the λ *int* gene, rather than to the *Salmonella hin* gene and, like λ integration, Fim-mediated inversion depends on the host IHF protein (66). There are thus at least two families of specific invertase proteins, one related to the Tn*3* resolvase, and a second related to λ integrase.

3.5.4 Additional inversion systems

The DNAs of two naturally occurring resistance plasmids, p15 and R64, exhibit complex patterns of DNA rearrangement. Each plasmid contains multiple, rather than just single, invertible segments because several related inversion sites can be used equivalently. Rearrangement in plasmid p15 is mediated by a Hin-related protein, whereas inversion in R64 is mediated by an Int-related protein. Additional DNA inversion systems related to the Hin system of *Salmonella* have been found in *Moraxella bovis* (controlling pilin synthesis), and in a penicillinase plasmid of *Staphylococcus aureus* (61).

In summary, phase variation due to DNA inversion is widespread. The relatedness of some invertase genes to Tn*3* resolvase and others to λ integrase are in accord with the idea that these genes have been spread by interspecific gene transfer and as yet unknown genome rearrangement processes. A second type of phase variation is caused by homologous recombination between divergent sequences of expressed and non-expressed genes (see Section 2.5). A third case of phase variation not discussed here is due to readily revertible frameshift mutations in a highly repetitive sequence (for review, see ref. 30). In each case, the resultant low level population heterogeneity should enhance survival in specific sets of variable and possibly hostile environments.

3.6 DNA excision and nitrogen fixation in cyanobacteria

Certain bacterial species can undergo terminal differentiation that formally resembles somatic cell differentiation in higher animals. We consider in this and the next section two cases associated with DNA rearrangement.

The filamentous green alga *Anabena* sp. derives energy by photosynthesis and also fixes atmospheric nitrogen. However, because oxygen produced during photosynthesis inactivates enzymes needed for nitrogen fixation, these two activities occur in different cell types. Depletion of the supply of reduced nitrogen induces the conversion of some cells in each filament to nitrogen-fixing 'heterocysts' that are incapable of further cell division. Heterocyst formation is associated with two precise 'excision' events: the removal of an 11 kb DNA segment at 11 bp direct repeats from within a silent *nifD* gene (which after excision encodes a subunit of the nitrogenase complex), and the removal of a 55 kb segment at 5 bp direct repeats from within a similarly interrupted ferredoxin gene (67).

The 11 kb segment and adjacent sequences have been cloned, and excision has been detected at low frequency in *E.coli*. A gene (*xisA*) whose product mediates excision of the 11 kb segment in *E.coli* has been identified by transposon insertion mutagenesis. A test involving transplacement of the mutant *xisA* allele into the *Anabena* genome showed that *xisA* is needed for excision in *Anabena* as well. Excision of the 55 kb segment occurred normally in *xisA⁻ Anabena*, however, indicating that different sets of genes probably control each excision event (68).

3.7 DNA rearrangement that changes transcription specificity

Stress-tolerant endospores are formed by *Bacillus subtilis* during nutrient limitation. This involves first the differentiation of a sporangium and then its partition into a forespore and a spore mother cell. The latter lyses after producing a protein shell which encases the mature spore. Terminal differentiation of the spore mother cell is associated with the appearance of a new sigma subunit of RNA polymerase, and a new pattern of transcription initiated at promoters used only during spore mother cell formation. The new sigma factor is encoded by two separate and incomplete genes that are expressed only after being spliced together at specific 5 bp direct repeats during mother cell development. A gene that is likely to encode the specific recombinase has been found among sporulation-defective mutations that block the rearrangement (69). Further tests will show whether this step in terminal differentiation of the spore mother cell is due to an excision event as in *Anabena*, inversion, or some other rearrangement process.

4. Transposition

4.1 Overview

Transposable elements are discrete DNA segments that move to new genomic locations without extensive sequence homology. They generate insertion mutations and extensive genome rearrangements, and alter the expression of nearby genes (70). This movement (transposition) is mediated by element-encoded transposase proteins, often acting in

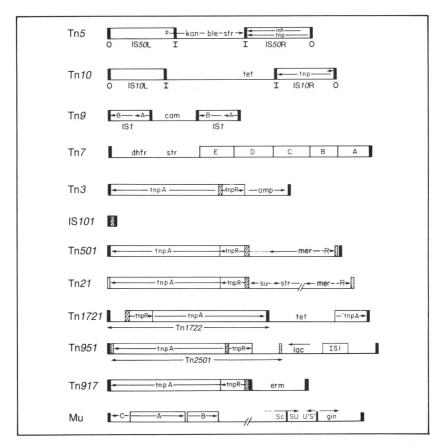

Figure 8. Representative bacterial transposable elements. Tn5, Tn10, and Tn9 are representative composite transposons, containing terminal repeats of insertion sequences. Tn3, IS101, Tn501, Tn21, Tn1721, and Tn951 from Gram-negative bacteria, and Tn917 from Gram-positive bacteria are related. (Adapted from ref. 72.)

concert with host factors on short sequences at the element ends. Transposition is generally independent of functions needed for generalized recombination. Transposable elements are diverse, and found in all groups of organisms. In bacteria they constitute an important evolutionary force, by helping spread genes encoding antibiotic resistance, virulence, and other traits among unrelated species, and by promoting various rearrangements within individual bacterial lineages (71). They have also attracted great attention as potent tools for genetic analysis and manipulation of many bacterial species (72).

Some representative prokaryotic elements are diagrammed in *Figure 8*. Insertion sequences such as IS*1* and IS*50* are the simplest of these elements; they contain just the genes and sites needed for their transposition and are generally less than 2 kb long. Transposons are generally larger and more complex, containing genes for auxiliary traits such as antibiotic resistance as well as genes for transposition proteins. Some of these transposons are composites, containing terminal inverted

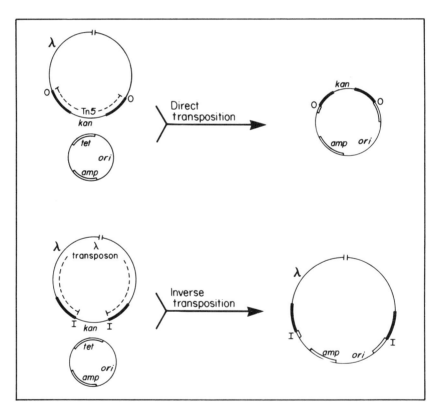

Figure 9. Comparison of direct and inverse transposition involving a λ::Tn*5* phage and a pBR322 target. Transposition mediated by the 'outside' (O) ends inserts Tn*5* into pBR322. Transposition mediated by the 'inside' (I) ends inserts λ into pBR322. (Reprinted from ref. 73, with permission.)

or direct repeats of IS elements bracketing auxiliary genes [e.g. Tn*5* (73), Tn*9* and Tn*1681* (74), and Tn*10* (75) in *Figure 8*]. Their mobility stems simply from the ability of a pair of IS elements to move in unison and carry along interstitial segments.

Figure 9 illustrates some of the flexibility conferred by composite transposons using an example of Tn*5* and its component IS*50* elements. Either end of the IS elements can participate in transposition. In the case of a λ phage carrying Tn*5*, use of the 'outside' (O) ends of IS*50* elements

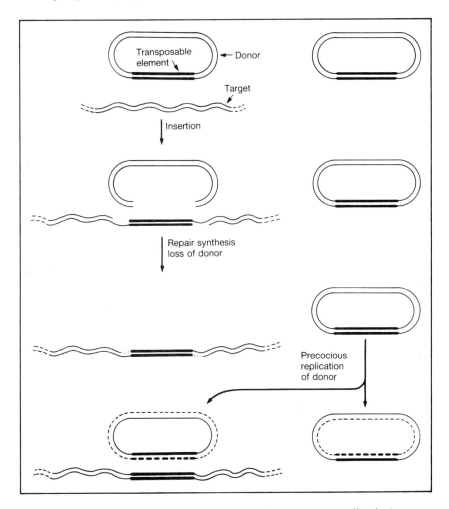

Figure 10. Model for conservative (cut-and-paste, non-replicative) transposition. Thick lines, transposable element; wavy lines, target; broken lines, newly synthesized DNA. The linear remnant of the DNA that had served as the transposition donor does not survive; a sibling of the donor molecule replicates out-of-turn to fill the niche created by this loss. As a result, duplicate copies of the element are found in the generations following transposition: one at the original site, and a new one at the transposed site. (Reprinted from ref. 70, with permission.)

causes transposition of the kanamycin (*kan*) resistance segment (Tn*5*) from λ to the target DNA; the reciprocal use of the IS*50* 'inside' (I) ends causes transposition of the entire λ genome away from the *kan* segment to the target (76,77). A similar test with a chromosomal insertion of Tn*10* demonstrated transposition of the entire bacterial chromosome into a target DNA molecule using the inside ends of IS*10* elements (78).

Other transposons differ in functional organization. Tn*3* (52) and Tn*7* (79) do not contain smaller complete IS elements. Tn*1721*, a Tn*3*-related tetracycline-resistance transposon, contains a 'minor transposon' at one end and an incomplete copy of this same element as a direct repeat at the other end. Cells carrying Tn*1721* can become resistant to very high levels of tetracycline by homologous recombination between the direct repeats and amplification of the *tet* genes (52,80). Tn*501* and Tn*951*, which are also in the Tn*3* family, carry IS elements within them which are unrelated to the transposon ends and (unlike the minor transposon of Tn*1721*) do not participate in movement of the entire transposon (52).

Yet another structural class of transposable elements is represented by temperate phage Mu (41,81). This transposon contains genes for all aspects of lytic growth and lysogeny in addition to the genes for transposition (A and B). Mu DNA is replicated during lytic growth only by transposition, and each replication event can cause a large scale DNA rearrangement as detailed below.

4.2 Transposition mechanisms

There are at least two classes of transposition mechanisms, conservative or cut-and-paste (82,83), and replicative (84). In conservative transposition (*Figure 10*) the element is cleaved from vector sequences, moved to a target site without net replication, and the linear remnant of vector DNA made during this process is lost rather than recircularized. It is generally believed that transposons Tn*5* and Tn*10* and their component IS elements move by this kind of cut-and-paste mechanism (73,75,82,83,85). The ability of conservative events to increase transposable element copy numbers reflects, in part, the way in which replication in bacteria is controlled: when an additional copy of donor DNA (that had not participated in transposition) is present in the cell it is likely to over-replicate to fill the resulting niche (*Figure 10*). In addition, transposition from one arm of a replication fork to a target site that had not yet been replicated (86) should also increase the copy number of the transposable element relative to other genomic sequences (*Figure 11*).

In replicative transposition (*Figure 12*) single-strand cleavages are made between element and vector DNA, and the free ends of the element are joined to target DNAs while still attached to the vector (84). The element is then copied, generating a 'co-integrate' in which donor and target DNAs are joined by direct repeats of the element. Tn*3* and related elements transpose in this way (52), as does phage Mu during lytic growth (81),

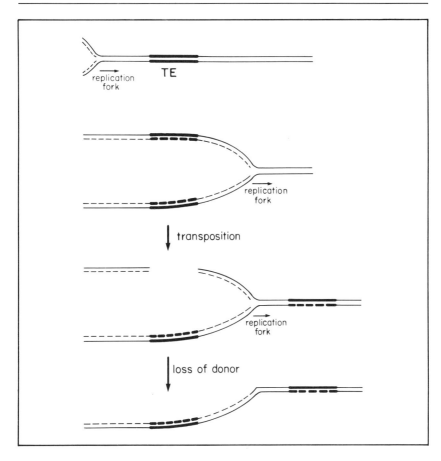

Figure 11. Intramolecular conservative transposition from one arm of a partially replicated DNA molecule (adapted from ref. 86). The sequences left in the arm of the replication fork that had served as the transposition donor are lost, perhaps by exonucleolytic degradation. The other arm of the replication fork persists and, as in *Figure 10*, in subsequent generations the element is found at both new and old sites. TE; transposable element.

and apparently also IS*1* in a fraction of transposition events (74,87).

The Tn*3*-related element, γδ, is a normal component of the F factor of *E.coli*; replicative transposition of γδ generates Hfr subclones in F⁺ cultures (7,88). This mode of Hfr formation is complementary to that involving *recA*-dependent recombination between homologous sequences in F and in the chromosome (Section 2). F also mediates the transfer of 'non-conjugative' plasmids such as pBR322 to recipient cells following γδ transposition and the formation of F – pBR322 co-integrates (89). Co-integrates made by transposition of γδ and other Tn*3*-related elements are short-lived: they are broken down to complementary donor and target DNAs, each containing a copy of the element, by a potent site-specific recombination reaction (resolution; Section 3).

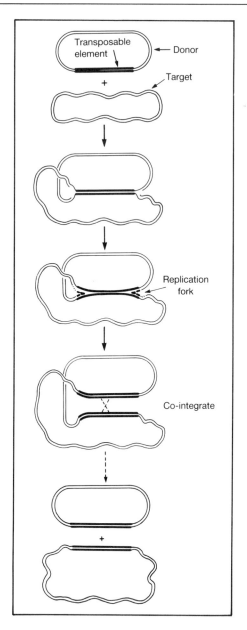

Figure 12.   Replicative intermolecular transposition (adapted from ref. 97).

4.3 Insertion specificity

The different transposable elements vary greatly in their insertion specificity. At one extreme Mu seems random when insertions in intervals of just a few kilobases are mapped (90), although insertion seems to be much more frequent in some intervals than in others (72,81). At the other extreme Tn*554* of Gram-positive bacteria and Tn*7* of Gram-negative bacteria each insert at high frequency into just a single site in the chromosomes of their hosts (79,91), a specificity reminiscent of that of phage λ integration discussed above.

Transposon Tn*10* inserts preferentially into just one or two sites per gene. The most active of these hotspots match the 6 bp consensus sequence GCTNAGC at five or six positions (92). IS*1* inserts preferentially into very AT-rich regions (93). The most frequently used region in pBR322 contains a site for IHF binding, and IHF also binds IS*1* ends. Interactions involving IHF and perhaps other proteins bound to IS*1* ends and to the target DNA may thus contribute to IS*1* insertion specificity (94).

Tn*5* can insert into multiple (perhaps 100) sites in typical genes, a few of which are used preferentially. GC pairs are present at each end of the 9 bp duplicated by insertion at most insertion hotspots, and are important for efficient insertion (95,96). Tn*5* inserts only into negatively supercoiled DNA. Recent tests have indicated that high supercoiling is needed for insertion at some hotspots, but not at others (J.K.Lodge and D.E.Berg, in preparation). This complex pattern may reflect quantitative contributions from several different factors to the chance of insertion at any given site. This relatively relaxed insertion specificity, coupled with an ability to transpose in many bacterial species, has led to the wide use of Tn*5* in molecular genetics (72,97).

4.4 Consequences of intramolecular transposition

In cases of transposition within a single DNA molecule, both conservative and replicative mechanisms generate rearrangements (*Figures 13* and *14*). During conservative transposition involving a composite element, when the transposon and target site are in one alignment, transposition inverts the DNA segment on one side of the target site relative to that on the other, and also changes the orientation of the two IS elements (from inverted to direct repeats in the case diagrammed here) (*Figure 13*, left panel). When the transposon and target site are in the other alignment (right panel) two complementary deleted daughter molecules are formed, each with a copy of the element. A linear fragment that corresponds to the central portion of the parental transposon is also generated, does not circularize, and is lost.

Intramolecular replicative transposition (*Figure 14*) also causes rearrangements: inversions when the element and target site are in one orientation (left panel) and two daughter molecules, each with a copy of the element when the element and target are in the other orientation (right panel). As one illustration, F′ plasmids and chromosomes with complementary deletions are formed by intramolecular transposition of γδ in Hfr strains (19). In a second example, lytic Mu growth or partial induction of Mu in a lysogen results in subgenomic circular DNAs containing a copy of Mu, and also large inversions in the chromosome,

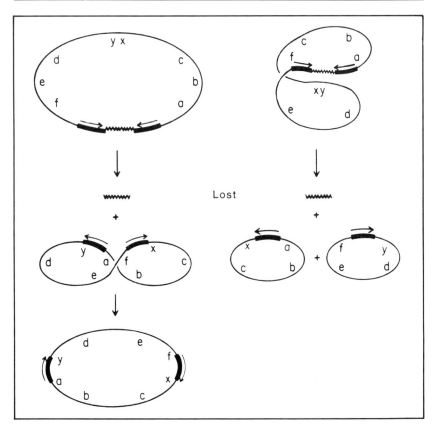

Figure 13. Deletion and inversion products formed by intramolecular conservative transposition. Transposition is to the x–y target site with the resulting structures depending on the alignment of the target and transposable element. a–f are hypothetical gene loci. (Reprinted from ref. 72, with permission.)

each due to replicative transposition. Further transposition generates more complex rearrangements, including inserts of subgenomic fragments bracketed by Mu into other chromosomal and plasmid DNAs (98). This ability of Mu to generate rearrangements has been exploited with specially engineered Mu vectors to permit cloning of *E. coli* DNAs *in vivo* (72,97–99).

Intramolecular IS*1* transposition in Tn*9*-containing plasmids, that may involve either the conservative or the replicative mechanism, has been used to generate nested deletions to facilitate DNA sequencing (100).

4.5 Regulation of transposition

Studies of Tn*3*, Mu, Tn*5*, and Tn*10* have shown that the mobility of transposable elements is regulated by a variety of mechanisms. Most allow

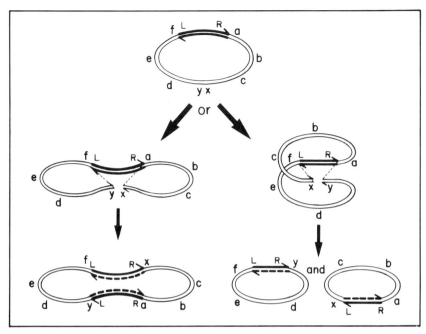

Figure 14. Deletion and inversion products formed by intramolecular replicative transposition. L and R indicate the left and right ends of the transposon. See *Figure 13* legend for other definitions of nomenclature. (Reprinted from ref. 72, with permission.)

an element to transpose more frequently when invading a cell lacking homologous elements rather than after it has become established. This would help disperse transposable elements while minimizing potentially deleterious transpositions and rearrangements. Several mechanisms link the probability of transposition to normal DNA replication, whch can be important for elements that transpose by a break – join (non-replicative) mechanism.

4.5.1 Transcriptional control of transposition: Mu and Tn3

Mu transposition is controlled by the prophage repressor. In established lysogens, the repressor blocks the expression of the A and B transposition genes, as well as phage lytic functions. Inactivation of the repressor, for example using a thermo-inducible prophage, permits expression of the A and B genes and multiple rounds of transposition. Partial or transient induction of a Mu lysogen can result in limited transcription of A and B, a few cycles of transposition rearrangement of host sequences, and then restoration of a stable prophage state (81,97).

Tn*3* transposition is also controlled by a protein repressor, in this case, the protein that also acts as resolvase. The promoter for transcription of the transposase gene is within the 120 bp *res* site, and resolvase, binding to this site, blocks access of RNA polymerase to the promoter. The repression of Tn*3* transposition is leaky by comparison with that of Mu, in part because the resolvase (*tnpR*) gene promoter is also within the *res* site, and resolvase thus autoregulates its own synthesis. Dilution of resolvase by cell growth may allow bursts of synthesis of transposase and resolvase until repression is again strengthened (52).

   A variant type of transcriptional control is exhibited by the Tn*3*-related mercury-resistance transposon Tn*501* (*Figure 8*). In this element, transcription of the transposase (*tnpA*) and resolvase (*tnpR*) genes are controlled by the repressor of the mercury-resistance operon. Low levels of mercury salts, which inactivate the repressor and induce expression of the resistance genes, also induce expression of *tnpA* and *tnpR*, and thereby promote Tn*501* transposition and the resolution of co-integrates (52).

4.5.2 Cis action of transposase

The transposase proteins of Tn*5*, Tn*9*, and Tn*10* act preferentially in *cis*, and when several copies of the same element are present the transposase acts preferentially on the DNA segment that has encoded it (101,102). As a result, the element that is most active transcriptionally is most prone to transpose. Localized action of transposase can be attributed to DNA binding by the nascent transposase protein, or a dependence on a high local concentration of the protein for activity.

4.5.3 Translational control of transposase synthesis

The synthesis of the Tn*10*(IS*10*) transposase is controlled in part by a repressor of translation. The repressor is a short mRNA molecule complementary to the 5′ end of the *tnp* mRNA. Base-pairing blocks ribosome loading and also leads to irreversible inactivation of the mRNA by creating an RNase III cleavage site. This RNA repressor is *trans*-acting and its effectiveness increases strongly with IS*10* copy number (75,103).

   An additional translational control mechanism, operating in the cases of both IS*50* and IS*10*, ensures that transposase synthesis can be controlled autogenously, essentially unaffected by changes in the transcription of adjacent host sequences. Sequences upstream of the normal transcriptional start constitute an inverted repeat of the sequence at the ribosome binding site for transposase. When transcription from an outside promoter leads to *tnp* mRNA containing these sequences, intramolecular base-pairing can block ribosome loading and the synthesis of transposase protein (104,105). This mechanism helps keep transposition subject to element-specific controls and relatively immune to the transcriptional activity of the locus into which the element has inserted.

   Transcription across the ends of IS*1*, IS*10*, and IS*50* also directly decreases their activity as transposition substrates (104–107). This might be due to displacement of transposition proteins, or to the RNA polymerase unwinding the DNA.

4.5.4 Post-translational inhibition of transposition

Tn5 transposition is regulated by a *trans*-acting inhibitor protein that interferes with the action, not the synthesis, of transposase. The inhibitor is encoded in the same reading frame as transposase from a translational start 55 codons downstream from the start of transposase synthesis. Most of the inhibitor is made from a separate promoter adjacent to the transposase promoter, and from an mRNA which starts within the open reading frame for transposase (108,109). As detailed below, this permits separate control of transposase and inhibitor synthesis.

4.5.5 Linking transposition to DNA synthesis: Dam methylation and DnaA protein

Dam methylation inhibits the transposition of IS*10*- and IS*50*-related elements in two ways. The promoters for transcription of the *tnp* genes contain GATC sites and are more active in Dam⁻ (methylation-deficient) than in Dam⁺ (methylation-proficient) cells (108–110). Methylation occurs slowly after DNA replication. Direct tests in the case of IS*10* have shown that one of the two hemi-methylated species of *tnp* promoter is about as active as the non-methylated promoter (110). The inside (I) ends of both IS*10* and IS*50* also contain GATC sites. These ends are more active as substrates for transposition proteins in Dam⁻ than in Dam⁺ cells, and also more active when hemi-, rather than fully methylated (109–111). Thus a dependence on Dam methylation helps coordinate the time of increased availability of one end of both IS*10* and IS*50* to the burst of expression of their cognate transposase genes, and helps link the movement of these elements to DNA replication.

The outside (O) end of IS*50* contains a binding site for DnaA protein, and DnaA protein is needed for Tn5(IS*50*) transposition (112–114). The DnaA protein is also needed to initiate bacterial chromosome replication: DnaA activity is modulated by the binding of ATP (active) versus ADP (inactive) which, in turn, is probably controlled physiologically (115). The requirement of Tn5 for DnaA protein may further link transposition to active DNA replication.

Linkage of transposition to replication is likely to be important for elements such as IS*50* and IS*10* which move by a conservative mechanism. It should help ensure that multiple copies of a donor sequence will be present in the cell and, consequently, that loss of the DNA remnant from which transposition had occurred will not cause cell death or loss of the donor DNA molecule from the cell lineage (*Figures 10* and *11*).

4.5.6 Transposition immunity

The ability of Tn*3*-related elements to insert into specific target DNA molecules can be regulated by sequences within the targets themselves, a phenomenon termed transposition immunity. Studies with small plasmids have shown that the presence of Tn*3* or just the 38 bp from one

Tn*3* end (a transposase binding site) blocks further Tn*3* insertion into that molecule (116).

Transposition immunity is very specific: an end of a divergent member of the Tn*3* family does not interfere with Tn*3* insertion into the target, and in some cases actually makes the nearby region highly preferred for insertion (117). This novel form of regulation may be due to sequestration of a limiting level of transposase by the Tn*3*-containing target in a form that makes this transposase unavailable to potential donor molecules (52). It seems advantageous in decreasing the chance of two separate Tn*3* insertions into the same molecule, thus minimizing the risk of deletion of part of this target molecule by homologous recombination or site-specific resolution.

In summary, the movement of transposable elements can be regulated in diverse ways: transcription or translation of the transposase gene; the activity of the transposase product; the ability of the element ends to act as transposition substrates; or the ability of other DNAs to serve as transposition targets.

5. Illegitimate recombination

Illegitimate recombination can be defined operationally as the joining of DNA sequences with little or no homology, resulting in the formation of deletions, duplications, fusions of separate DNA molecules, and other rearrangements (118–122). It is distinct from the recombination and site-specific recombination discussed above and is *recA*-independent in some but not all test systems (118,122–128). The study of illegitimate recombination reflects an intrinsic interest in mutation processes and DNA–protein interactions, their contributions to somatic and germ line mutations in humans, the utility of illegitimate recombination in bacterial genetics and, conversely, the complications it can cause for recombinant DNA analyses. The evidence reviewed below indicates that illegitimate recombination can occur by several different mechanisms, some involving DNA breakage without replication, and others involving errors during DNA synthesis.

5.1 Rearrangements associated with DNA breakage

Two types of rearrangements have been identified that seem to result from specific DNA breakage without net replication.

5.1.1 Restriction endonuclease-mediated rearrangements

'Restriction endonucleases' are sometimes found in bacterial species that seem not to restrict foreign (non-modified) DNAs, which suggests that they might mediate DNA sequence rearrangements *in vivo* (129). *In vivo* deletion events that were strongly stimulated by *Eco*RI were demonstrated using a chloramphenicol (*cam*) resistance gene that had been inactivated by insertion at its *Eco*RI site. Reversion from sensitivity to resistance to chloramphenicol was stimulated at least 1000-fold by use of cells containing the gene that encodes *Eco*RI. The Cam-resistant revertant plasmids obtained were identical in structure to those generated by *in vitro* removal of the insert at the *Eco*RI site, and re-ligation (130).

5.1.2 DNA gyrase-mediated rearrangements

E.coli DNA gyrase breaks double-stranded DNA, passes a DNA helix through the break, and then reseals the broken DNA ends in a concerted reaction that can increase the negative superhelical density of circular DNAs (131). One intermediate in this reaction consists of linear DNA, with gyrase bound to the ends, held in a circular configuration by the interaction between subunits of the active gyrase protein.

In vitro studies using λ and pBR322-type DNAs showed that DNA gyrase can fuse two unrelated DNAs at low frequencies. The fusions occur preferentially at sequences that correspond to sites of gyrase action, and duplications or deletions are often present at the site of fusion. This reaction is stimulated by oxolinic acid (a gyrase inhibitor which traps the linear DNA – gyrase intermediates) (132 – 134). Equivalent illegitimate recombination reactions have been found using T4-encoded DNA gyrase

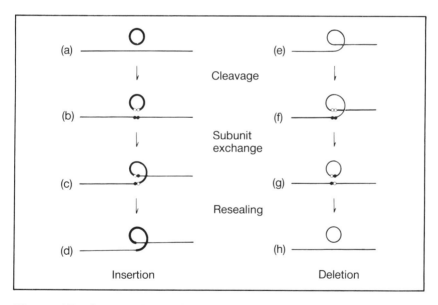

Figure 15. A model for DNA gyrase-mediated illegitimate recombination. (a – d) Fusion of two different DNAs. (e – h) Deletion within a single DNA molecule. Double-stranded DNAs are represented by lines. Circles bound to DNAs represent gyrase subunits. The circles are drawn open and filled to emphasize how the exchange of gyrase protein subunits would mediate this exchange. (Reprinted from ref. 132, with permission.)

(135). *In vivo*, λ-pBR322 fusions formed in *recA⁻* cells closely resemble
the rearrangements formed using DNA gyrase *in vitro* (136). These
rearrangements may result from alignment and recombination of subunits
of a tetrameric gyrase protein at the stage in a supercoiling reaction
between DNA cleavage and re-ligation (*Figure 15*).

Studies of duplications have shown that the 'repetitive extragenic
palindromic' (REP) sequences (1000 copies per genome; 137) also provide
preferred endpoints for DNA sequence rearrangements (tandem
duplications in the case studied) (G.F.-L.Ames, personal communication).
These REP sequences have been shown recently to be sites of DNA
gyrase binding (138). Experiments in progress will test whether
rearrangements involving REP sequences result from the action of DNA
gyrase, or by other mechanisms, for example involving their repetitive
or palindromic sequences (G.F.-L.Ames, personal communication).

5.1.3 Linear DNA transformation and deletion

Circular deleted plasmids are recovered at low frequency after
transformation of *E.coli* cells with linearized pBR322-related DNAs. DNA
sequencing showed that most deletion endpoints occur in short direct
repeats, and that the recovery of viable plasmids depends on the *xth*
(exonuclease III) gene. These deleted plasmids may therefore arise by
limited exonucleolytic erosion of the ends after transformation *in vivo*,
annealing of complementary sequences, and then repair synthesis and
ligation (139). This implies that some deletions may result from double-
strand breakage and then equivalent erosion of the ends, repair, and
ligation *in vivo*.

5.2 Deletions associated with initiation and termination of replication

A single DNA strand is synthesized from double-stranded template DNAs
of coliphage M13, and of plasmid pC194 (from Gram⁺ bacteria), and also
from conjugative F factor plasmids during DNA transfer. Single-strand
synthesis is initiated from the 3'OH end of a parental strand formed by
an origin-specific single-strand nick. The origin and terminus of replication
coincide and the initiator protein also helps terminate replication by
cleaving the completed single strand and then joining 5' and 3' ends (140).
Studies of M13 (141,142), and of M13-based chimeric plasmids in which
deletions can be selected directly (143) showed that deletions end
preferentially at the M13 replication origin/terminus, and that their
formation depends on the M13-specific initiator protein (gene II protein)
(143).

The other or 'variable' deletion endpoints are clustered in just a few
preferred regions, some resembling the origin/terminus in sequence, and
some coinciding with other palindromic sequences. These deletions may
result from initiation of replication at the correct site and then incorrect

termination. Considerable complexity in this deletion process was indicated by the finding that the use of one subset of preferred endpoints depends on the *E.coli rep* gene (which specifies a helicase needed for M13 replication fork movement), whereas the use of another subset is impaired by depletion of ligase (using a temperature-sensitive mutant), and the use of a third subset of endpoints is independent of both *rep* and ligase activities (122).

Members of a distinctive class of F' plasmids (that do not arise by homologous recombination or by intramolecular transposition) have one deletion endpoint at the origin for conjugative DNA transfer, and another endpoint at variable positions in the first DNA transferred early by their Hfr parents. These F' plasmids may arise during conjugation by a deletion mechanism similar to that invoked for M13, above: normal initiation of Hfr transfer, copying of the transferred DNA strand in the recipient cell, and then incorrect termination of the copying reaction (144).

Another pattern of preferred deletion endpoints has been found with plasmid pC194. Some deletions in this plasmid seem to result from incorrect initiation, but correct termination, whereas others resemble those in M13 and F in resulting from correct initiation but incorrect termination of replication (122,145).

5.3 Illegitimate recombination associated with direct and inverted repeats

Most deletions, and perhaps other rearrangements as well, seem (from their sequences) not to be caused by DNA gyrase, restriction endonucleases, or other known endonucleases, nor to be due to errors at the initiation or termination of replication. Many of them end in direct repeats of $5-10$ bp, and/or are associated with short, imperfect palindromes. Additional studies summarized below indicate that 'excision' of transposons Tn*5* and Tn*10* probably results from similar deletion mechanisms, and that tests based on reversion of insertion mutations can be used to understand deletion processes.

5.3.1 Short direct and inverted repeats and deletion formation

A number of direct repeats in the range of $5-10$ bp occur by chance in any typical gene. Although these sequence matches are too short for classical recombination they do serve as hotspots for deletion formation. This was first shown by Miller and colleagues (127), who found that about half of a set of deletions ranging up to several hundred base pairs in size in the *E.coli lacI* gene ended in direct repeats of $5-8$ bp. Among longer $(700-1000$ bp) deletions, two-thirds ended in a hyphenated 14/17 bp direct repeat, and most of the rest ended in direct repeats of at least 4 bp. Point mutations which changed one repeat in the hotspot from a 14/17 to a 13/17 bp match decrease the frequency of deletions at this hotspot (128). Many subsequent experiments, using a variety of different plasmid, chromosomal, and phage replicons, support the conclusion that direct repeats constitute preferred sites for deletion formation (reviewed in refs 121,122).

An involvement of direct repeats is in accord with one of two models.

(i)   Slippage of nascent and template DNA strands between the first and second copy of the repeat could result in a failure to copy interstitial sequences (124,127,128,146).

(ii)   Alternatively, deletions may result from DNA breakage *in vivo*, limited erosion of the ends, and joining at direct repeats, equivalent to that invoked in transformation with linearized plasmid DNAs (124,147).

Discussion of tests of these models is deferred until Section 5.3.7.

Deletion endpoints are also often associated with short imperfect inverted repeats (palindromes), in some cases just a few bp in length (128,147). In one model, palindromes might stimulate deletion formation by intrastrand pairing of complementary sequences to form a hairpin structure, which would then be cleaved by conformation-specific endonucleases (147). Alternatively, hairpin structures might impede DNA synthesis and thereby promote slippage of the growing DNA strand (127,128,148,149).

5.3.2 Direct repeats and tandem duplications

Two studies have shown that short direct repeats are also important in the formation of tandem duplications. A derivative of *E.coli* with higher levels of expression of a chromosomally encoded β-lactamase was found to result from tandem duplication and then further amplification of a 10 kb segment containing the *ampC* gene. The duplication endpoints coincided with 12 bp direct repeats (150). Similarly, the endpoints of most (28/30) duplications of the *lac* operon also occurred in short direct repeats, most frequently in a 14/18 bp direct repeat (151).

5.3.3 Intermolecular exchange at direct repeats

Short matched sequences have been implicated in the fusion of different DNA molecules. Three of four co-integrates generated by the fusion of λ and pBR322 in *recA*⁺ cells resulted from joining at 10 or 11 bp micro-homologies, and the fourth apparently resulted from two joining events in micro-homologies of 10 and 13 bp (152). Similarly, in *Bacillus subtilis*, fusions between two different plasmids, and between a plasmid and the bacterial chromosome, usually involved short sequence matches (153,154).

5.3.4 Transposon excision: Tn5 and Tn10

Mutations due to Tn*5* and Tn*10* insertions revert at frequencies ranging from less than 10^{-10} up to about 10^{-4}, depending on the precise site of insertion and whether the insert is in the chromosome or a plasmid. Reversion entails deletion of the transposon plus one copy of the 9 bp target duplication that was generated during insertion. Transposon 'excision' is thus at least formally equivalent to spontaneous deletion processes. Additional studies have shown that:

(i)   excision is transposase-independent, and not correlated with transposition to new sites;

(ii) the excision frequency varies over a greater than 1000-fold range at different insertion sites in the same gene, and is about 100-fold more frequent from F′ plasmids than from identical sites in the bacterial chromosome; transposition frequencies, in contrast, are not much affected by such differences in location;

(iii) shortening of the terminal inverted repeats, or conversion of the long inverted repeats to direct repeats at the transposon ends, markedly decreases the frequencies of excision without affecting transposition (148,149,155).

Tn5 and Tn10 are polar on the expression of distal genes in an operon. An equivalent deletion phenomenon, sometimes termed imprecise or nearly precise excision, can be detected by selecting for relief of this polarity. Typically, either the element and adjacent sequences are deleted, or most of the element is deleted, but part of it remains at the site of insertion. Imprecise excision is also transposase-independent, and the deletion endpoints also tend to coincide with short direct repeats (123,149).

Each of these results indicates that the excision of transposons Tn5 and Tn10 results from spontaneous deletion events. This realization, and the ease of making insertion mutations, manipulating the insert sequences, and assaying their reversion, have made the transposon-derived inserts good material for analyses of illegitimate recombination processes.

Transposon excision, like other spontaneous deletion events, has generally been ascribed to errors in DNA synthesis. Hairpin structures formed by intrastrand pairing in palindromic template DNA would foster slippage by both impeding elongation of nascent DNA strands, and by juxtaposing the flanking 9 bp direct repeats. The end of a nascent strand stalled at the first copy of the 9 bp repeat could branch migrate and anneal with the second copy of the direct repeat, and again prime DNA synthesis. The resultant new DNA strand would have the ancestral (revertant) DNA sequence (124,148,149). Formally, deletion might also be due to DNA cleavage at transposon ends (perhaps caused by extrusion of hairpin structures when the DNA is highly supercoiled), exonucleolytic erosion, complementary base-pairing at the direct repeats, and then re-ligation of the broken ends, as in the case of transformation with linearized plasmid DNA (see Section 5.1.3) (124). Finally, a model invoking recombination proteins specific for transposable elements has also been proposed (156).

A variety of E.coli mutations that alter the excision frequencies of Tn5 and Tn10 have been isolated. Mutations that increase excision were found in dam, mutH, mutL, and mutS (adenine methylation and methyl-directed mismatch repair), ssb (single-stranded DNA-binding protein), mutD (dnaQ; replication fidelity), recB, recC (DNA unwinding; exo- and endonucleolytic activity; and recombination) (125,126), and *uup*, a previously unknown gene (157). Some of these mutations did not enhance the deletion of a 50 bp remnant of Tn*10* left after imprecise excision, suggesting that different factors can limit the deletion of small and large DNA segments. The stimulation by most mutations was *recA*-independent (as is the basal, unstimulated frequency of excision in many test systems); the stimulation by one *recB* allele was, however, *recA*-dependent (126). The enhanced deletion frequencies caused by three various mutations might reflect stimulation of a normal deletion pathway or the creation of new pathways.

Other evidence bearing on deletion mechanisms came from findings that Tn*5* excision is about 100-fold more frequent from F' plasmids than from the same sites in the bacterial chromosome.

(i)   A test with *recA* partial diploid strains containing distinguishable Tn*5* elements at the same site in chromosomal and F' *lacZ* genes showed that excision is about 50-fold more frequent from the plasmid than from the chromosome, and that this stimulation by F depends on the physical linkage of Tn*5* to F DNA (155).

(ii) Conjugation tests showed that the frequency of excision was higher from F' plasmids that had been transferred than from those that had remained in the donor cell (155).

(iii) A mutation in F that stimulated excision also stimulated F' transfer, while a mutation which eliminated this stimulation rendered the F' plasmid defective in transfer (155,156).

Because single DNA strands are transferred in bacterial conjugation, we proposed that F-enhanced Tn*5* excision reflects the formation of deletions due to errors in copying recently transferred single-stranded DNAs (155).

There may be additional *trans* effects in *recA*+ cells. The frequency of reversion of insertion mutations in various chromosomal genes was found to be higher in *recA*+ F' strains than in related F− strains (156,158). This was correlated with more frequent recombination between long repeated sequences in pBR322 plasmid, and was explained as a physiological response to F' DNA transfer: the single-strandedness of transferred DNA might induce an SOS response which includes error-prone DNA synthesis (repair) and recombination processes (158). Thus, F may stimulate Tn*5* excision in *trans* as well as in *cis*, reflecting the action of proteins that increase errors during DNA synthesis and that are inducible in *recA*+ strains (*trans*) (158), as well as errors in the copying of single DNA strands transferred by conjugation to *recA*+ or *recA*− cells (*cis*) (155).

5.3.5 Phage Mu excision

Bacteriophage Mu does not contain long terminal inverted repeats equivalent to those in Tn*5* and Tn*10*, and selection for reversion or the relief of polarity indicated that fully repressed Mu prophages do not undergo excision (frequencies of $<10^{-10}$). Mu prophages can undergo excision, however, if they are derepressed, and they contain a mutation in Mu gene B to prevent high frequencies of transposition and host cell killing. Further tests showed that Mu excision, unlike Tn5 and Tn10 excision, depends on the Mu transposase (gene A product) (159,160), which is reminiscent of the transposase-dependent excision of several eukaryotic elements (see ref. 70). To better understand this process, we selected products of Mu excision by relief of polarity and sequenced them. Most contained simple deletions with endpoints that coincided with short direct repeats. Several had a more complex rearrangement, however: one end of Mu transposed to a new site, coincident with deletion of most of Mu, leaving in some cases an inverted repeat of remaining Mu sequences. In one model, formation of simple deletions is due to the binding of the A protein to the Mu ends, which might promote DNA cleavage or strand slippage during normal replication. The more complex rearrangement might arise by an abortive replicative transposition event and then template switching of the nascent DNA strand (161), a model similar to that invoked earlier for λdv formation (Section 5.4.2).

5.3.6 Local DNA sequence effects on deletion specificity

Tests based on the reversion of insertion mutations may be generally useful as a complement to the traditional approach of sequencing many deletion endpoints since:

(i)   *a priori* knowledge of the location of the endpoints of deletions leading to reversion eliminates the need to sequence each deletion

(ii)   rare as well as frequent events can be easily detected and measured with precision, simply by counting revertant colonies

(iii)   identical DNA segments can be placed throughout a gene and changed at will, thereby permitting systematic examination of parameters affecting deletion events

(iv)   sequences so unstable that they would have been lost from the genome during evolution can often be generated as insertion mutations and analyzed using reversion tests.

We have used reversion tests to evaluate more closely the effects of inverted and direct repeat sequences and other DNA sequence parameters on deletion formation. In one experiment, a series of identical palindromes ranging from 22–90 bp in size were derived from inserts of Tn5-related elements inserted at different sites in the *amp* gene of pBR322. The frequency of deletion of the 22 bp palindromic insert plus one copy of the flanking 9 bp direct repeat (made during transposon insertion) varied over a 100-fold range (10^{-8} to 10^{-6}). Lengthening of the palindrome to 90 bp increased the deletion frequency, but to a variable extent (8-fold to 18 000-fold) depending on the site (124). To better assess the importance of direct repeats during the deletion of palindromic sequences, the lengths of direct repeats of *amp* gene sequence bracketing several palindromic inserts were changed systematically in pBR322. Amp[R] revertants were not found (frequency $<10^{-10}$) with direct repeats of 2 or 4 bp. Elongation of the direct repeats from $7-11$ bp increased the frequency of deletion about 10^4–fold (from 5×10^{-10} and 2×10^{-9} with the 22 and 90 bp palindromes, respectively) (W.-Y.Chow, D.Warnock, and D.E.Berg, unpublished data).

These studies confirmed that both palindromes and direct repeats strongly stimulate deletion formation. They also indicated that other factors, in addition to direct and inverted repeat length, must also affect the process (124). One of these additional factors appears to be local GC content (162).

5.3.7 Deletion by copy choice (replication error)

The formation of deletions during replication was shown unequivocally using pBR322 derivatives containing an origin for M13 single-stranded DNA synthesis and also insertions of a Tn10-related element in the *amp* or *tet* genes. M13 replication entails first the synthesis of one single DNA strand, and then its complement. The onset of M13-specific replication of these plasmids caused the frequency of cells carrying revertants (deletants) to increase rapidly from 10^{-6} to nearly 100% of the population. Physical studies showed that the first single strand was copied with fidelity from the double-stranded template, and that deletions were formed in the next step during the copying of the extended single strand into duplex DNA. There was no DNA transfer of the type predicted by break–join models between parental (non-deleted) and progeny (deleted) molecules. These results showed that deletions of palindromic DNA sequences can arise during the replication of single-stranded DNA (163).

Chromosomal and most other plasmid DNAs do not have a persistent single-stranded phase, however, which made it important to test whether they undergo deletion by the same mechanism. Evidence that deletion in pBR322 also involves slipped mispairing during replication has been obtained by analyses of deletions that can end in either of two sets of adjacent direct repeats, and tests of how the choice of these endpoints is affected by the length of the palindromic segment undergoing deletion (164). One model for deletion by slipped mispairing, and the stimulation of deletion by palindromic sequences is shown in *Figure 16*.

5.4 Duplications and deletions in phage λ

Two examples from earlier phage λ literature provide important unsolved problems that should now be amenable to molecular genetic analysis.

5.4.1 Tandem duplications in phage λ

Tandem duplications of λ DNA proved relatively easy to select because of two characteristics of λ:

(i)   The length of DNA packaged into a λ virion is determined by the positions of *cos* (cohesive end, or DNA packaging) sites.

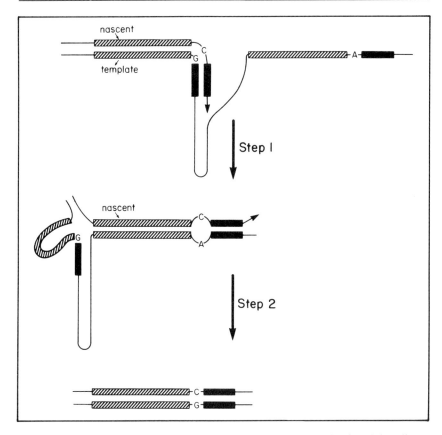

Figure 16. A slipped mispairing model for deletion formation involving direct and inverted repeats. Two different direct repeats are shown: one pair (hatched boxes) brackets a palindromic segment; a second pair (filled boxes) is displaced by 1 bp, and one component is within the palindromic sequence. In the model, DNA synthesis is impeded by hairpin structures in single-stranded template DNA, which stimulates slippage (branch migration) of the end of the nascent DNA strand. The particular case shown here indicates the reversion of a series of insertion mutations in the *amp* gene of pBR322. The frequency of deletion endpoints in the repeats indicated by filled boxes relative to those indicated by hatched boxes increased with palindrome length. This provided evidence that deletions in plasmid pBR322 result from errors in replication rather than DNA breakage and erosion of the ends. (Reprinted from ref. 164, with permission.)

(ii)   A minimum genome length of about 37 kb (75% of wild-type λ) is needed for phage viability.

Tandem duplications were selected as large plaque mutants in stocks of phage λ whose genome size is suboptimal for growth, or by density in the CsCl equilibrium gradient (reflecting the increased DNA content and DNA:protein ratio of the duplication phage). Electron microscope heteroduplex mapping showed that the duplication endpoints were widely distributed, and studies with mutants indicated that the formation of duplications is independent of known site-specific and generalized recombination functions (165,166). Many duplications formed in cells infected with two different genetically marked phage contained alleles from both parents, even though defects in *red* and *rec* (phage and bacterial recombination) genes had eliminated normal crossing-over (167). The duplications were found at frequencies of about 10^{-3} in stocks of a phage strain called λ*tdel*33 (a deletion variant of a λ/φ80 hybrid), and 10^{-5} in λ*b*221 (a simple deletion derivative of phage λ). Finally, mixed infections showed that λ*tdel*33 can stimulate the formation of duplications in λ*b*221 (165–167).

Such tandem duplications would seem to provide a powerful means of analyzing illegitimate recombination processes. If, as assumed, these phage result from duplications in monomeric λ DNAs, do they arise by intermolecular *rec*-independent breakage and joining of two different molecules, or by the switching of templates during DNA replication? Alternatively, might these duplication phage actually arise in dimeric or large DNAs, formed during replication or by the joining of different λ DNA ends during infection? In this case they might arise by simple deletion (e.g. as in *Figure 16* or *17*), rather than by a duplication mechanism.

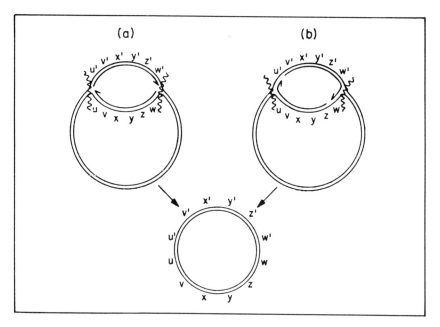

Figure 17.   Models for the formation of λ*dv* deletion mutant derivatives of phage λ consisting of inverted repeats. (a) and (b) represent bidirectionally replicating λ DNA molecules with different strand configurations at the growing point. In (a) a break-join event precedes abnormal replication; in (b) abnormal replication (the switching of templates, formally akin to the slippage diagrammed in *Figure 16*) precedes DNA breakage and joining. (Reprinted from ref. 168, with permission.)

What is the nature of the stimulating factor specified by λ*tdel*33? Does it actually operate in *trans*, or only in *cis* on DNAs physically linked to λ*tdel*33?

5.4.2 Massive deletions in phage λ

λ*dv* plasmids are deletion derivatives of phage λ that can retain as little as 5 – 8% of the original λ genome, just the genes needed for autonomous and controlled replication (11,12,16). Analyses of λ*dv* formation offered a special opportunity to examine deletion events involving nearly an entire DNA molecule. Genetic and electron microscopic characterizations showed that many λ*dv* plasmids are formed by simple deletion of a contiguous segment of the λ chromosome. In some cases, however, the deletion process was also associated with simultaneous duplication and inversion of most or all of the remaining λ sequences. The relative yields of inverted dimer and simple deletion types of structures depended on the genotype of the parental phage. We postulated (12, 168) that the inverted duplications could arise in partially replicated molecules either by template switching or DNA breakage at the replication fork (*Figure 17*).

6. Acknowledgements

This work was supported by grants from the National Science Foundation (DMB-8608193), the United States Public Health Service (GM-37138), and the Lucille P.Markey Charitable Trust.

7. References

1. Riley, M. and Krawiec, S. (1987) Genome organization. In *Escherichia Coli and Salmonella Typhimurium: Cellular and Molecular Biology*. Neidhardt, F. C., Ingraham, J. C., Low, K. B., Magasanik, B., Schaechter, M., and Umbarger, H. E. (eds). American Society for Microbiology, Washington DC, p. 967.
2. Galas, D. J. and Chandler, M. (1989) Bacterial insertion sequences. In *Mobile DNA*. Berg, D. E. and Howe, M. M. (eds). American Society for Microbiology, Washington DC, p. 109.
3. Noller, H. F. and Nomura, M. (1987) Ribosomes. In *Escherichia Coli and Salmonella Typhimurium: Cellular and Molecular Biology*. Neidhardt, F. C., Ingraham, J. C., Low, K. B., Magasanik, B., Schaechter, M., and Umbarger, H. E. (eds). American Society for Microbiology, Washington DC, p. 104.
4. Low, K. B. (1988) *The Recombination of Genetic Material*. Academic Press, New York.
5. Kucherlapati, R. and Smith, G. R. (1988) *Genetic Recombination*. American Society for Microbiology, Washington DC.
6. Deonier, R. C. (1987) Locations of native insertion sequence elements. In *Escherichia Coli and Salmonella Typhimurium: Cellular and Molecular Biology*. Neidhardt, F. C., Ingraham, J. C., Low, K. B., Magasanik, B., and Umbarger, H. E. (eds). American Society for Microbiology, Washington DC, p. 982.
7. Low, K. B. (1987) Hfr strains of *Escherichia Coli* K-12. In *Escherichia Coli and Salmonella Typhimurium: Cellular and Molecular Biology*. Neidhardt, F. C., Ingraham, J. C., Low,

K. B., Magasanik, B., Schaechter, M., and Umbarger, H. E. (eds). American Society for Microbiology, Washington DC, p. 1134.

8. Seed, B. (1983) Purification of genomic sequences from bacteriophage libraries by recombination and selection *in vivo*. *Nucleic Acids Res., 11*, 2427.

9. Shen, P. and Huang, H. V. (1986) Homologous recombination in *Escherichia coli*: dependence on substrate length and homology. *Genetics, 112*, 441.

10. Balbas, P., Soberon, X., Merino, E., Zurita, M., Lomell, H., Valle, F., Flores, N., and Bolivar, F. (1986) Plasmid vector pBR322 and its special purpose derivatives—a review. *Gene, 50*, 3.

11. Matsubara, K. and Kaiser, A. D. (1968) λ*dv*: an autonomously replicating DNA fragment. *Cold Spring Harbor Symp. Quant. Biol., 33*, 769.

12. Berg, D. E. (1974) Genetic evidence for two types of gene arrangements in new λ*dv* plasmid mutants. *J. Mol. Biol., 86*, 59.

13. Kellenberger-Gujer, G., Boy de l Tour, E., and Berg, D. E. (1974) Transfer of the λ*dv* plasmid to new bacterial hosts. *Virology, 58*, 576.

14. Bedbrook, J. R. and Ausubel, F. M. (1976) Recombination between bacterial plasmids leading to the formation of plasmid multimers. *Cell, 9*, 707.

15. Hobom, G. and Hogness, D. S. (1974) The role of recombination in the formation of circular oligomers of the λ*dv*1 plasmid. *J. Mol. Biol., 88*, 65.

16. Berg, D. E. (1971) Regulation in phage with duplications of the immunity region. In *The Bacteriophage Lambda*. Hershey,A.D. (ed.). Cold Spring Harbor Laboratory, Cold Spring Harbor, NY, p. 667.

17. Berg, C. M., Liu, L., Coon, M., Strausbaugh, L., Gray, P., Vartak, N. B., Brown, M., Talbot, D., and Berg, D. E. (1989) pBR322-derived multicopy plasmids harboring large inserts are often dimers in *Escherichia coli* K-12. *Plasmid, 21*, 138.

18. Berg, C. M. and Curtiss, R. III (1967) Transposition derivatives of an Hfr strain of *Escherichia coli* K12. *Genetics., 56*, 503.

19. Holloway, B. and Low, K. B. (1987) F-prime and R-prime factors. In *Escherichia Coli and Salmonella Typhimurium: Cellular and Molecular Biology*. Neidhardt, F. C., Ingraham, J. C., Low, K. B., Magasanik, B., Schaechter, M., and Umbarger, H. E. (eds). American Society for Microbiology, Washington DC, p. 1145.

20. Anderson, P. and Roth, J. (1981) Spontaneous tandem genetic duplications in *Salmonella typhimurium* arise by unequal recombination between rRNA (*rrn*) cistrons. *Proc. Natl. Acad. Sci. USA, 78*, 3113.

21. Konrad, E. B. (1969) The genetics of chromosomal duplications. Ph.D. Thesis, Harvard University, Cambridge, MA.

22. Konrad, E. B. (1977) Method for the isolation of *Escherichia coli* mutants with enhanced recombination between chromosomal duplications. *J. Bacteriol., 130*, 167.

23. Schmid, M. B. and Roth, J. R. (1983) Selection and endpoint distribution of bacterial inversion mutations. *Genetics, 105*, 539.

24. Francois, V., Louarn, J., Patte, J., and Louarn, J. -M. (1987) A system for *in vivo* selection of genomic rearrangements with predetermined endpoints in *Escherichia coli* using modified Tn*10* transposons. *Gene, 56*, 99.

25. Segall, A., Mahan, M. J., and Roth, J. R. (1988) Rearrangement of the bacterial chromosome: forbidden inversions. *Science, 241*, 1314.

26. Rebollo, J. -E., Francois, V., and Louarn, J. -M. (1988) Deletion and possible role of two large nondivisible zones on the *Escherichia coli* chromosome. *Proc. Natl. Acad. Sci. USA, 85*, 9391.

27. Marinus, M. G. (1987) Methylation of DNA. In *Escherichia Coli and Salmonella Typhimurium: Cellular and Molecular Biology*. Neidhardt, F. C., Ingraham, J. C., Low, K. B., Magasanik, B., Schaechter, M., and Umbarger, H. E. (eds). American Society for Microbiology, Washington DC, p. 697.

28. Sternberg, N. (1985) Evidence that adenine methylation influences DNA – protein interactions in *Escherichia coli*. *J. Bacteriol., 164*, 490.

29. Brewer, B. J. (1988) When polymerases collide: replication and the transcriptional organization of the *E.coli* chromosome. *Cell, 53*, 679.

30. Swanson, J. and Koomey, J. M. (1989) Mechanisms for variation of pili and outer membrane protein II in *Neisseria gonorrhoeae*. In *Mobile DNA*. Berg, D. E. and Howe, M. M. (eds). American Society for Microbiology, Washington DC, p. 743.

31. Meyer, T. F., Miawer, N., and So, M. (1982) Pilus expression in *Neisseria gonorrhoeae* involves chromosomal rearrangement. *Cell, 30*, 45.

32. Koomey, M., Gotschlich, E. C., Robbins, K., Bergstrom, S., and Swanson, J. (1987) Effects of *recA* mutations on pilus antigenic variation and phase transitions in *Neisseria gonorrhoeae*. *Genetics,* **117**, 391.
33. Seifert, H. S., Ajioka, R. S., Marchal, C., Sparling, P. F., and So, M. (1988) DNA transformation leads to pilin antigenic variation in *Neisseria gonorrhoeae*. *Nature,* **336**, 392.
34. Echols, H. and Guarneros, G. (1983) Control of integration and excision. In *Lambda II*. Hendrix, R. W., Roberts, J. W., Stahl, F. W., and Weisberg, R. A. (eds). Cold Spring Harbor Laboratory, Cold Spring Harbor, NY, p. 75.
35. Thompson, J. F. and Landy, A. (1989) Regulation of bacteriophage lambda site-specific recombination. In *Mobile DNA*. Berg, D. E. and Howe, M. M. (eds). American Society for Microbiology, Washington DC, p. 1.
36. Campbell, A. (1971) Genetic structure. In *The Bacteriophage Lambda*. Hershey, A. D. (ed.). Cold Spring Harbor Laboratory, Cold Spring Harbor, NY, p. 13.
37. Simon, M. N., Davis, R. W., and Davidson, N. (1971) Heteroduplexes of DNA molecules of lambdoid phages: physical mapping of their base sequence relationships by electron microscopy. In *The Bacteriophage Lambda*. Hershey,A.D. (ed.). Cold Spring Harbor Laboratory, Cold Spring Harbor, NY, p. 313.
38. Campbell, A. and Botstein, D. (1983) Evolution of the lambdoid phages. In *Lambda II*. Hendrix, R. W., Roberts, J. W., Stahl, F. W., and Weisberg, R. A. (eds). Cold Spring Harbor Laboratory, Cold Spring Harbor, NY, p. 365.
39. Arber, W. (1983) A beginner's guide to lambda biology. In *Lambda II*. Hendrix, R. W., Roberts, J. W., Stahl, F. W., and Weisberg, R. A. (eds). Cold Spring Harbor Laboratory, Cold Spring Harbor, NY, p. 381.
40. Yarmolinsky, M. B. and Sternberg, N. (1988) Bacteriophage P1. In *The Bacteriophages*. Calendar,R. (ed.). Plenum, NY, p. 291.
41. Symonds, N., Toussaint, A., van de Putte, P., and Howe, M. M. (1987) *Phage Mu*. Cold Spring Harbor Laboratory, Cold Spring Harbor, NY.
42. Bertani, L. E. and Six, E. W. (1988) The P2-like phages and their parasite, P4. In *The Bacteriophages*. Calendar,R. (ed.). Plenum Press, NY, p. 73.
43. Hershey, A. D. and Dove, W. (1971) Introduction to lambda. In *The Bacteriophage Lambda*. Hershey, A. D. (ed.). Cold Spring Harbor Laboratory, Cold Spring Harbor, NY, p. 3.
44. Friedman, D. I. and Gottesman, M. (1983) Lytic mode of lambda development. In *Lambda II*. Hendrix, R. W., Roberts, J. W., Stahl, F. W., and Weisberg, R. A. (ed). Cold Spring Harbor Laboratory, Cold Spring Harbor, NY, p. 21.
45. Gottesman, M. E. and Weisberg, R. A. (1971) Prophage insertion and excision. In *The Bacteriophage Lambda*. Hershey, A. D. (ed.). Cold Spring Harbor Laboratory, Cold Spring Harbor, NY, p. 113.
46. Freifelder, D. and Levine, E. E. (1975) The formation of polylysogens during infection of *Escherichia coli* with bacteriophage λ. *Virology,* **63**, 428.
47. Shimada, K., Weisberg, R. A., and Gottesman, M. E. (1972) Prophage λ at unusual locations. I. Locations of the secondary attachment sites and the properties of the lysogens. *J. Mol. Biol.,* **63**, 483.
48. Shimada, K., Weisberg, R. A., and Gottesman, M. E. (1973) Prophage λ at unusual chromosomal locations. II. Mutations induced by bacteriophage λ in *E.coli* K-12. *J. Mol. Biol.,* **80**, 297.
49. Schrenk, W. J. and Weisberg, R. A. (1975) A simple method for making new transducing lines of coliphage lambda. *Mol. Gen. Genet.,* **137**, 101.
50. Nash, H. A. (1975) Integrative recombination of bacteriophage lambda DNA *in vitro*. *Proc. Natl. Acad. Sci. USA,* **72**, 1072.
51. Hatfull, G. F. and Grindley, N. D. F. (1988) Resolvase and DNA-invertases: a family of enzymes active in site-specific recombination. In *Genetic Recombination*. Kucherlapati, R. and Smith, G. R. (eds). American Society for Microbiology, Washington DC, p. 357.
52. Sherratt, D. J. (1989) Tn*3* and related transposable elements: Site-specific recombination and transposition. In *Mobile DNA*. Berg, D. E. and Howe, M. M. (eds). American Society for Microbiology, Washington DC, p. 163.
53. Summers, D. K. and Sherratt, D. J. (1984) Multimerization of high copy number plasmids causes instability: colE1 encodes a determinant essential for plasmid monomerization and stability. *Cell,* **36**, 1097.

54. Stirling, C. J., Stewart, G., and Sherratt, D. J. (1988) *Mol. Gen. Genet.*, **214**, 80.
55. Austin, S., Ziese, M., and Sternberg, N. (1981) A novel role for site-specific recombination in maintenance of bacterial replicons. *Cell*, **25**, 729–736.
56. Sternberg, N., Sauer, B., Hoess, R., and Abremski, K. (1986) An initial characterization of the bacteriophage P1 *cre* structural gene and its regulatory region. *J. Mol. Biol.*, **187**, 197.
57. Hoess, R. H., Ziese, M., and Sternberg, N. (1982) P1 site-specific recombination: nucleotide sequence of the recombining sites. *Proc. Natl. Acad. Sci. USA*, **79**, 3398.
58. Hoess, R. H., Wierzbicki, A., and Abremski, K. (1986) The role of the *loxP* spacer region in P1 site-specific recombination. *Nucleic Acids Res.*, **14**, 2287.
59. Chesney, R. H., Scott, J. R., and Vapnek, D. (1979) Integration of the plasmid prophages P1 and P7 into the chromosome of *Escherichia coli. J. Mol. Biol.*, **130**, 161.
60. Sternberg, N., Hamilton, D., and Hoess, R. (1981) P1 site-specific recombination. II. Recombination between *loxP* and the bacterial chromosome. *J. Mol. Biol.*, **150**, 487.
61. Glasgow, A. C., Hughes, K. T., and Simon, M. I. (1989) Bacterial DNA inversion systems. In *Mobile DNA*. Berg, D. E. and Howe, M. M (eds). American Society for Microbiology, Washington DC, p. 637.
62. Zieg, J., Silverman, M., Hilmen, M., and Simon, M. (1977) Recombinational switch for gene expression. *Science*, **196**, 170.
63. Silverman, M. and Simon, M. (1983) Phase variation and related systems. In *Mobile Genetic Elements*. Shapiro,J.A. (ed.). Academic Press, NY, p. 537.
64. Koch, C., Mertens, G., Rudt, F., Kahmann, R., Kanaar, R., Plasterk, R. A., van de Putte, P., Sandulache, R., and Kamp, D. (1987) The invertible G segment. In *Phage Mu*. Symonds, N., Toussaint, A., van de Putte, P., and Howe,M.M. (eds). Cold Spring Harbor Laboratory, Cold Spring Harbor, NY, p. 75.
65. Klemm, P. (1986) Two regulatory *fim* genes, *fimB* and *fimE* control the phase variation of type I fimbriae in *Escherichia coli. EMBO J.*, **5**, 1389.
66. Eisenstein, B. I., Sweet, D., Vaughn, V., and Friedman, D. I. (1987) Integration host factor is required for the DNA inversion that controls phase variation in *Escherichia coli. Proc. Natl. Acad. Sci. USA*, **84**, 6506.
67. Haselkorn, R. (1989) Excision of elements interrupting nitrogen fixation operons in cyanobacteria. In *Mobile DNA*. Berg,D.E. and Howe,M.M. (eds). American Society for Microbiology, Washington DC, p. 735.
68. Golden, J. W. and Wiest, D. R. (1988) Genome rearrangement and nitrogen fixation blocked by inactivation of *xisA* gene. *Science*, **242**, 1421.
69. Stragier, P., Kunkel, B., Kroos, L., and Losick, R. (1989) Chromosomal rearrangement generating a composite gene for a developmental transcription factor. *Science*, **243**, 507.
70. Berg, D. E. and Howe, M. M. (1989) *Mobile DNA*. American Society for Microbiology, Washington DC.
71. Berg, D. E. (1989) Transposable elements in prokaryotes. In *Gene Transfer in the Environment*. Levy, S. B. and Miller, R. (eds). McGraw-Hill, NY, p. 99.
72. Berg, C. M., Berg, D. E., and Groisman, E. A. (1989) Transposable elements and the genetic engineering of bacteria. In *Mobile DNA*. Berg,D.E. and Howe,M.M. (eds). American Society for Microbiology, Washington DC, p. 879.
73. Berg, D. E. (1989) Transposon Tn*5*. In *Mobile DNA*. Berg, D. E. and Howe, M. M. (eds). American Society for Microbiology, Washington DC, p. 185.
74. Galas, D. J. and Chandler, M. (1989) Bacterial insertion sequences. In *Mobile DNA*. Berg, D. E. and Howe, M. M. (eds). American Society for Microbiology, Washington DC, 109.
75. Kleckner, N. (1989) Transposon Tn*10*. In *Mobile DNA*. Berg, D. E. and Howe, M. M. (eds). American Society for Microbiology, Washington DC, p. 227.
76. Sasakawa, C. and Berg, D. E. (1982) IS*50* mediated inverse transposition: discrimination between the two ends of an IS element. *J. Mol. Biol.*, **159**, 257.
77. Nag, D. K., DasGupta, U., Adelt, G., and Berg, D. E. (1985) IS*50*-mediated inverse transposition: specificity and precision. *Gene*, **34**, 17.
78. Harayama, S., Oguchi, T., and Iino, T. (1984) The *E.coli* K-12 chromosome flanked by two IS*10* sequences transposes. *Mol. Gen. Genet.*, **197**, 62.
79. Craig, N. L. (1989) Transposon Tn*7*. In *Mobile DNA*. Berg, D. E. and Howe, M. M. (eds). American Society for Microbiology, Washington DC, p. 211.

80. Schmitt, R., Altenbuchner, J., Wiebauer, K., Arnold, W., Puhler, A., and Schoffl, A. (1989) Basis of transposition and gene amplification by Tn*1721* and related tetracycline-resistance transposons. *Cold Spring Harbor Symp. Quant. Biol.,* **40**, 59.
81. Pato, M. L. (1989) Bacteriophage Mu. In *Mobile DNA.* Berg, D. E. and Howe, M. M. (eds). American Society for Microbiology, Washington DC, p. 23.
82. Berg, D. E. (1977) Insertion and excision of the transposable kanamycin resistance determinant Tn*5.* In *Insertion Elements, Plasmids and Episomes.* Bukhari, A. I., Shapiro, J. A., and Adhya, S. L. (eds). Cold Spring Harbor Laboratory, Cold Spring Harbor, NY, p. 205.
83. Berg, D. E. (1983) Structural requirement for IS*50*-mediated gene transposition. *Proc. Natl. Acad. Sci. USA,* **80**, 792.
84. Shapiro, J. A. (1979) Molecular model for the transposition and replication of bacteriophage Mu and other transposable elements. *Proc. Natl. Acad. Sci. USA,* **76**, 1933.
85. Bender, J. and Kleckner, N. (1986) Genetic evidence that Tn*10* transposes by a nonreplicative mechanism. *Cell,* **45**, 801.
86. Greenblatt, I. M. and Brink, R. A. (1963) Transposition of modulator in maize into divided and undivided chromosome segments. *Nature,* **197**, 412.
87. Biel, S. W. and Berg, D. E. (1984) Mechanism of IS*1* transposition in *E.coli*: choice between simple insertion and cointegration. *Genetics,* **108**, 319.
88. Guyer, M., Reed, R. R., Steitz, J. A., and Low, K. B. (1981) Identification of a sex factor affinity site in *E.coli* as γδ. *Cold Spring Harbor Symp. Quant. Biol.,* **45**, 135.
89. Guyer, M. (1978) The γδ sequence of F is an insertion sequence. *J. Mol. Biol.,* **126**, 347.
90. Bukhari, A. I. and Zipser, D. (1972) Random insertion of Mu-1 DNA within a single gene. *Nature New Biol.,* **236**, 240.
91. Murphy, E. (1989) Transposable elements in Gram-positive bacteria. In *Mobile DNA.* Berg, D. E. and Howe, M. M. (eds). American Society for Microbiology, Washington, DC, p. 269.
92. Halling, S. M. and Kleckner, N. (1982) A symmetrical six-base-pair target site determines Tn*10* insertion specificity. *Cell,* **28**, 153.
93. Zerbib, D., Gamas, P., Chandler, M., Prentki, P., Bass, S., and Galas, D. J. (1985) Specifity of insertion of IS*1. J. Mol. Biol.,* **185**, 517.
94. Prentki, P., Chandler, M., and Galas, D. J. (1987) *Escherichia coli* integration host factor bends the DNA at the ends of IS*1* and in an insertion hotspot with multiple IHF binding sites. *EMBO J.,* **6**, 2479.
95. Berg, D. E., Schmandt, M. A., and Lowe, J. B. (1983) Specificity of transposon Tn*5* insertion. *Genetics,* **105**, 813.
96. Lodge, J. K., Weston-Hafer, K., and Berg, D. E. (1988) Transposon Tn*5* target specificity: preference for insertion at GC pairs. *Genetics,* **120**, 645.
97. Berg, C. M. and Berg, D. E. (1987) Uses of transposable elements and maps of known insertions. In *Escherichia Coli and Salmonella Typhimurium: Cellular and Molecular Biology.* Neidhardt, F. C., Ingraham, J. C., Low, K. B., Magasanik, B., Schaechter, M., and Umbarger,H.E (eds). American Society for Microbiology, Washington DC, p. 1071.
98. Toussaint, A. and Resbois, A. (1983) Phage Mu: transposition as a life-style. In *Mobile Genetic Elements.* Shapiro, J. A. (ed.). Academic Press, NY, p. 105.
99. Groisman, E. A. and Casadaban, M. J. (1986) Mini-Mu bacteriophage with plasmid replicons for *in vivo* cloning and *lac* gene fusing. *J. Bacteriol.,* **168**, 357.
100. Ahmed, A. (1987) Use of transposon-promoted deletions in DNA sequence analysis. *Methods Enzymol.,* **155**, 177.
101. Isberg, R. R. and Syvanen, M. (1981) Replicon fusion promoted by the inverted repeats of Tn*5.* The right repeat is an insertion sequence. *J. Mol. Biol.,* **150**, 15.
102. Morisato, D., Way, J. C., Kim, H. J., and Kleckner, N. (1983) Tn*10* transposase acts preferentially on nearby transposon ends *in vivo. Cell,* **32**, 799.
103. Simons, R. W. and Kleckner, N. (1983) Translational control of IS*10* transposition. *Cell,* **34**, 683.
104. Davis, M. A., Simons, R. A., and Kleckner, N. (1985) Tn*10* protects itself at two levels from fortuitous activation by external promoters. *Cell,* **43**, 379.
105. Sasakawa, C., Lowe, J. B., McDivitt, L., and Berg, D. E. (1982) Control of transposon Tn*5* transposition in *Escherichia coli. Proc. Natl. Acad. Sci. USA,* **79**, 7450.
106. Machida, C., Machida, Y., Wang, H. -C., Ishizaki, K., and Ohtsubo, E. (1983)

Repression of cointegration ability of insertion element IS*1* by transcriptional readthrough from flanking regions. *Cell,* **34**, 135.

107. Biel, S. W., Adelt, G., and Berg, D. E. (1984) Transcriptional control of IS*1* transposition in *Escherichia coli. J. Mol. Biol.,* **174**, 251.

108. Krebs, M. P. and Reznikoff, W. S. (1986) Transcriptional and translational initiation sites of IS*50*. Control of transposase and inhibitor expression. *J. Mol. Biol.,* **192**, 781.

109. Yin, J. C. P., Krebs, M. P., and Reznikoff, W. S. (1988) The effect of *dam* methylation on Tn*5* transposition. *J. Mol. Biol.,* **199**, 35.

110. Roberts, D., Hoopes, B. C., McClure, W. R., and Kleckner, N. (1985) IS*10* transposition is regulated by DNA adenine methylation. *Cell,* **43**, 117.

111. Dodson, K. W. and Berg, D. E. (1989) Factors affecting transposition activity of IS*50* and Tn*5* ends. *Gene,* **76**, 207.

112. Johnson, R. C. and Reznikoff, W. S. (1983) DNA sequences at the ends of transposon Tn*5* required for transposition. *Nature,* **204**, 280.

113. Phadnis, S. H. and Berg, D. E. (1987) Identification of base pairs in the IS*50* O end needed for IS*50* and Tn*5* transposition. *Proc. Natl. Acad. Sci. USA,* **84**, 9118.

114. Yin, J. C. P. and Reznikoff, W. S. (1987) *dnaA*, an essential host gene, and Tn*5* transposition. *J. Bacteriol.,* **169**, 4637.

115. Bramhill, D. and Kornberg, A. (1988) A model for initiation at origins of DNA replication. *Cell,* **54**, 915.

116. Lee, C. H., Bhagwat, A., and Heffron, F. (1983) Identification of a transposon Tn*3* sequence required for transposition immunity. *Proc. Natl. Acad. Sci. USA,* **80**, 6765.

117. Grinsted, J., Bennett, P. M., Higginson, S., and Richmond, M. H. (1978) Regional preference of insertion of Tn*501* and Tn*801* into RP1 and its derivatives. *Mol. Gen. Genet.,* **166**, 313.

118. Franklin, N. C. (1967) Extraordinary recombination events in *Escherichia coli*. Their independence of the *rec*+ function. *Genetics,* **55**, 699.

119. Weisberg, R. A. and Adhya, S. (1977) Illegitimate recombination in bacteria and bacteriophage. *Annu. Rev. Genet.,* **11**, 451.

120. Anderson, P. (1987) Twenty years of illegitimate recombination. *Genetics,* **115**, 581.

121. Allgood, N. D. and Silhavy, T. J. (1988) Illegitimate recombination in bacteria. In *Genetic Recombination*. Kucherlapati,R. and Smith,G.R. (eds). American Society of Microbiology, Washington DC, p. 309.

122. Ehrlich, S. D. (1989) Illegitimate recombination in bacteria. In *Mobile DNA*. Berg, D. E. and Howe, M. M. (eds). American Society of Microbiology, Washington DC, p. 799.

123. Collins, J., Volckaert, G., and Nevers, P. (1982) Precise and nearly precise excision of the symmetrical inverted repeats of Tn*5*: common features of *recA*-independent deletion events in *Escherichia coli. Gene,* **19**, 139.

124. DasGupta, U., Weston-Hafer, K., and Berg, D. E. (1987) Local DNA sequence control of deletion formation in *Escherichia coli* plasmid pBR322. *Genetics,* **115**, 41.

125. Lundblad, V. and Kleckner, N. (1984) Mismatch repair mutations in *Escherichia coli* K12 enhance transposon excision. *Genetics,* **109**, 3–19.

126. Lundblad, V., Taylor, A. F., Smith, G. R., and Kleckner, N. (1984) Unusual alleles of recB and recC stimulate excision of inverted repeat transposons Tn*5* and Tn*10*. *Proc. Natl. Acad. Sci. USA,* **81**, 824.

127. Farabaugh, P. I., Schmeissner, U., Hofer, M., and Miller, J. H. (1978) Genetic studies of the *lac* repressor. VII. On the molecular nature of spontaneous hotspots in the *lacI* gene of *Escherichia coli. J. Mol. Biol.,* **126**, 847.

128. Albertini, A. M., Hofer, M., Calos, M. P., and Miller, J. H. (1982) On the formation of spontaneous deletions: the importance of short sequence homologies in the generation of large deletions. *Cell,* **29**, 319.

129. Nathans, D. and Smith, H. O. (1975) Restriction endonucleases in the analysis and restructuring of DNA molecules. *Annu. Rev. Biochem.,* **44**, 273.

130. Chang, S. and Cohen, S. N. (1977) *In vivo* site-specific genetic recombination promoted by the *Eco*RI restriction endonuclease. *Proc. Natl. Acad. Sci. USA,* **74**, 4811.

131. Wang, J. C. (1985) DNA topoisomerases. *Annu. Rev. Biochem.,* **54**, 665.

132. Ikeda, H., Moriya, K., and Matsumoto, T. (1981) *In vitro* study of illegitimate recombination: involvement of DNA gyrase. *Cold Spring Harbor Symp. Quant. Biol.,* **45**, 399.

133. Ikeda, H., Aoki, K., and Naito, A. (1982) Illegitimate recombination mediated *in vitro*

by DNA gyrase of *Escherichia coli*: structure of recombinant DNA molecules. *Proc. Natl. Acad. Sci. USA,* **79**, 3724.

134. Ikeda, H., Kawasaki, I., and Gellert, M. (1984) Mechanism of illegitimate recombination: common sites for recombination and cleavage mediated by *E.coli* DNA gyrase. *Mol. Gen. Genet.,* **196**, 546.

135. Ikeda, H. (1986) Bacteriophage T4 DNA topoisomerase mediates illegitimate recombination *in vitro. Proc. Natl. Acad. Sci. USA,* **83**, 922.

136. Marvo, S. L., King, S. R., and Jaskunas, S. R. (1983) Role of short regions of homology in intermolecular illegitimate recombination events. *Proc. Natl. Acad. Sci. USA,* **80**, 2452.

137. Higgins, C. F., McLaren, R. S., and Newbury, S. F. (1988) Repetitive extragenic palindromic sequences, mRNA stability and gene expression: evolution by gene conversion?—a review. *Gene,* **7**, 3.

138. Yang, Y. and Ames, G. F. -L. (1988) DNA gyrase binds to the family of prokaryotic repetitive extragenic palindromic sequences. *Proc. Natl. Acad. Sci. USA,* **85**, 8850.

139. Conley, E. C., Saunders, V. A., Jackson, V., and Saunders, J. R. (1986) Mechanism of intramolecular recyclization and deletion formation following transformation of *Escherichia coli* with linearized plasmid DNA. *Nucleic Acids Res.,* **14**, 8919.

140. Kornberg, A. (1982) *Supplement to DNA Replication.* W. H. Freeman, San Francisco.

141. Griffith, J. and Kornberg,A. (1974) Mini M13 bacteriophage: circular fragments of M13 DNA are replicated and packaged during normal infections. *Virology,* **59**, 139.

142. Kornberg, A. (1980) *DNA Replication.* W.H. Freeman, San Francisco.

143. Michel, B. and Ehrlich, S. D. (1986) Illegitimate recombination at the replication origin of bacteriophage M13. *Proc. Natl. Acad. Sci. USA,* **83**, 3386.

144. Horowitz, B. and Deonier, R. C. (1985) Formation of delta-tra F' plasmids: specific recombination at *oriT. J. Mol. Biol.,* **186**, 267.

145. Michel, B. and Ehrlich, S. D. (1986) Illegitimate recombination occurs between the replication origin of the plasmid pC194 and a progressing replication fork. *EMBO J.,* **5**, 3691.

146. Streisinger, G., Okada, Y., Emrich, J., Newton, J., Tsugita, A., Terzhagi, E., and Inouye, M. (1966) Frameshift mutations and the genetic code. *Cold Spring Harbor Symp. Quant. Biol.,* **33**, 77.

147. Glickman, B. and Ripley, L. S. (1984) Structural intermediates in deletion mutagenesis: a role for palindromic DNA. *Proc. Natl. Acad. Sci. USA,* **81**, 512.

148. Egner, C. and Berg, D. E. (1981) Excision of transposon Tn5 is dependent on the inverted repeats but not on the transposase function of Tn5. *Proc. Natl. Acad. Sci. USA,* **78**, 459.

149. Foster, T. J., Lundblad, V., Hanley-Way, S., Halling, S. M., and Kleckner, N. (1981) Three Tn10-associated excision events: relationship to transposition and role of direct and inverted repeats. *Cell,* **23**, 215.

150. Edlund, T. and Normark, S. (1981) Recombination between short DNA homologies causes tandem duplication. *Nature,* **292**, 269.

151. Whoriskey, S. K., Nghiem, V. -H., Leong, P. -M., Masson, J. -M., and Miller, J. H. (1987) Gene amplification in *E.coli*: DNA sequences at the junctures of amplified gene fusions. *Genes Develop.,* **1**, 227.

152. King, S. R., Krolewski, M. A., Marvo, S. L., Lipson, P. J., Pogue-Geile, K. L., Chung, J. H., and Jaskunas, S. R. (1982) Nucleotide sequence analysis of *in vivo* recombinants between bacteriophage lambda DNA and pBR322. *Mol. Gen. Genet.,* **186**, 548.

153. Bashkirov, V. I., Stoilova-Disheva, M. M., and Prozorov, A. A. (1988) Interplasmidic illegitimate recombination in *Bacillus subtilis. Mol. Gen. Genet.,* **213**, 465.

154. Bashkirov, V. I., Khasanov, F. K., and Prozorov, A. A. (1988) Illegitimate recombination in *Bacillus subtilis*: nucleotide sequences at recombinant DNA junctions. *Mol. Gen. Genet.,* **210**, 578.

155. Berg, D. E., Egner, C., and Lowe, J. B. (1983) Mechanism of F factor enhanced excision of transposon Tn5. *Gene,* **22**, 1.

156. Hopkins, J. D., Clements, M. B., Liang, T. -Y., Isberg, R. R., and Syvanen, M. (1980) Recombination genes on the *Escherichia coli* sex factor specific for transposable elements. *Proc. Natl. Acad. Sci. USA,* **77**, 2814.

157. Hopkins, J. D., Clements, M., and Syvanen, M. (1983) New class of mutations in *Escherichia coli (uup)* that affect precise excision of insertion elements and bacteriophage Mu growth. *J. Bacteriol.,* **153**, 384.

158. Syvanen, M., Hopkins, J., Griffin, T., Liang, T., Ippen-Ihler, K., and Koloder, R. (1986) Stimulation of precise excision and recombination by conjugal proficient F plasmids. *Mol. Gen. Genet., 203*, 1.
159. Bukhari, A. I. (1975) Reversal of mutator phage Mu integration. *J. Mol. Biol., 96*, 87.
160. Khatoon, H. and Bukhari, A. I. (1981) DNA rearrangements associated with reversion of bacteriophage Mu induced mutations. *Genetics, 98*, 1.
161. Nag, D. K. and Berg, D. E. (1987) Specificity of bacteriophage Mu excision. *Mol. Gen. Genet., 207*, 395.
162. Singer, B. S. and Westlye, J. (1988) Deletion formation in bacteriophage T4. *J. Mol. Biol., 202*, 233.
163. Brunier, D., Michel, B., and Ehrlich, S. D. (1988) Copy choice illegitimate DNA recombination. *Cell, 52*, 883.
164. Weston-Hafer, K. and Berg, D. E. (1989) Palindromy and the location of deletion endpoints in *E.coli. Genetics, 121*, 651.
165. Emmons, S. W., MacCosham, V., and Baldwin, R. L. (1975) Tandem genetic duplications in phage lambda III. The frequency of duplication mutants in two derivatives of phage lambda is independent of known recombination systems. *J. Mol. Biol., 91*, 133.
166. Emmons, S. W. and Thomas, J. O. (1975) Tandem genetic duplications in phage lambda IV. The locations of spontaneously arising tandem duplications. *J. Mol. Biol., 91*, 147.
167. Emmons, S. W., MacCosham, V., and Baldwin, R. L. (1975) On the mechanism of production of tandem genetic duplications in phage lambda. *J. Mol. Biol., 95*, 83.
168. Chow, L. T., Davidson, N., and Berg, D. E. (1974) Electron microscope study of the structures of λdv DNAs. *J. Mol. Biol., 86*, 69.

2

Antigenic variation in African trypanosomes: genetic recombination and transcriptional control of VSG genes

Lex H.T.Van der Ploeg

1. Introduction

Diverse parasite species can escape immunodestruction since they parallel the repertoire of variable immunoglobulin genes of the host with a smaller repertoire of sequentially expressed antigens. Almost all of the mechanisms for differential control of parasite antigen gene expression leading to immune-evasion involve genetic recombination. Antigenic variation of the eukaryotic African trypanosomes and the prokaryote *Borrelia*, and pilus variation of the prokaryote *Neisseria* are examples of the unique adaptations of these organisms to their continuous surveillance by the host's immune system (1).

Other recent reviews discuss the molecular biology of trypanosomes (2–6), and present a comparison of the different modes of genetic recombination in eukaryotes (1) and the peculiarities of transcriptional control (7) and RNA processing (8,9) in trypanosomes. In this review, I will discuss the relevance of these topics to the mechanisms which control the genetic program underlying immune-evasion by the African trypanosomes.

2. African trypanosomes and immune-evasion

Trypanosomes are unicellular flagellated parasites. They are the causative agents of serious diseases such as sleeping sickness, caused by the African trypanosome *Trypanosoma brucei* and Chaga's disease caused by the Southern American trypanosome, *Trypanosoma cruzi*, in humans and their live-stock (10). The African trypanosomes have evolved an intricate mechanism which enables them to proliferate in the host's bloodstream. Individual cells periodically change the entire antigenic composition of their cell surface coat and thereby escape immune attack. Most of these parasites are transmitted from the bloodstream of one mammalian host to another by the tsetse fly which serves as an insect vector in which they undergo a complex developmental life-cycle. Exceptions to this mode of transmission can be seen in the trypanosome of horses, *Trypanosoma equiperdum*, which is venereally transmitted.

African trypanosomes belong to the taxonomic order of the Kineto-plastidae (11). All members of this order are characterized by a peculiar mitochondrion which differs from mitochondria in other eukaryotes. Kinetoplastidae mitochondria contain a DNA network of thousands of catenated, roughly 1.5 kb mini-circles and about 250 copies of a 23 kb maxi-circle. The maxi-circle represents the mitochondrial genome (12,13). This group of eukaryotic organisms is considered to be evolutionarily ancient, branching shortly after the division of archae- and eubacteria (14). Many of the peculiar and novel molecular mechanisms that have been identified in these protozoa are a reflection of their solitary evolutionary path.

2.1 The cell surface coat and antigenic variation

In the bloodstream of a mammalian host, the cell surface of the African trypanosome is covered by a dense protein coat which consists of about 10 million identical proteins known as the variant cell surface glycoproteins or VSGs (15 – 17). The VSGs are the predominant antigens of bloodstream-form trypanosomes. The VSG coat serves as an impermeable barrier against host antibodies and prevents recognition of any other common antigenic determinants that are concealed underneath the VSG coat. In addition, the VSG coat prevents the activation of the alternate complement pathway which leads to the lysis of naked trypanosomes (18).

The progression of a primary infection with *Trypanosoma brucei* (the causative agent of sleeping sickness) is schematically presented in *Figure 1* (19). The parasitemia is characterized by a steady increase in the number of parasites over about a 6 day period. The rising titer of IgM antibody directed against the parasite's cell surface coat finally leads to a decline in parasite numbers and eventually to the decimation of the population. However, by the end of the 6 day period, a small fraction of parasites will have spontaneously switched to the expression of an entirely different VSG protein coat. These organisms will escape immunodestruction and continue to proliferate in the bloodstream. Using immunological techniques, over 100 different cell surface VSG antigens have been identified in *T.brucei congolense* (20). In *T.brucei brucei*, the presence of over 1000 different VSG genes (including an unknown number of pseudogenes) has been estimated using molecular techniques (21). Given the large repertoire of VSG genes, the process of relapsing parasitemias can persist for a long period of time, and can sometimes result in the death of the host.

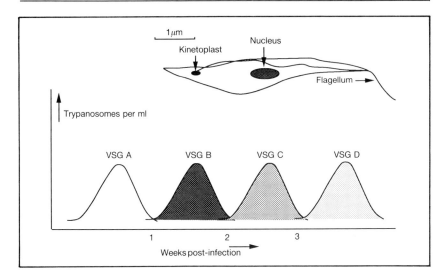

Figure 1. (Top) Schematic representation of a trypanosome and (bottom) the course of an infection with African trypanosomes. The figure represents a simplification of the situation *in vivo*, in which the trypanosome populations in each peak of the parasitemia are heterogeneous, and consist of several different variants. At peak parasitemia up to 10^9 parasites may exist per ml of blood.

Given the large repertoire of VSG genes, a nomenclature was developed to discriminate between different trypanosome clones (11,22). A trypanosome sample obtained from an infected animal or patient is referred to as a strain. Cloned trypanosomes from a strain are used to establish laboratory infections which are referred to as serodemes. The serodemes are named after the institute at which the serodeme was developed (e.g. MiTar; Molteno Institute Trypanosome antigen repertoire). Cloned trypanosomes from a particular serodeme are referred to as antigenic variants or variant antigen types (VATs). Individual variant antigen types have been given numbers for their identification (e.g. MiTat 1.2; Molteno Institute Trypanosome antigen type 1.2). These variant antigen types may also be referred to as variants, followed by a number which identifies the VSG (e.g. MiTat 1.2 is the same as variant 221 which expresses VSG 221).

2.2 Structure of the cell surface coat

The VSG coat functions as an impermeable barrier against antibodies, thus preventing the immunological recognition of common antigenic determinants which are concealed underneath the VSG coat. Several discoveries facilitated the elucidation of the structure of the VSG coat, explaining how it functions so efficiently in immune-evasion and in the concealment of common antigenic determinants.

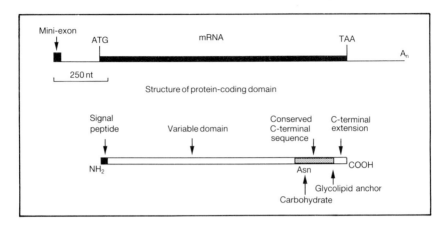

Figure 2.   VSG mRNA and the structure of the protein-coding domain.

(i)   The VSG is synthesized as a glycoprotein of roughly 60 kd, having
       an N-terminal signal peptide which is removed from the mature VSG
       (23). Significant sequence homologies between the different VSGs
       are confined to approximately 50 amino acids located at the C
       terminus and the remainder of the protein comprises the
       antigenically variable domain (*Figure 2*; 24–26). On the basis of
       sequence homologies and consensus cysteine residues at the C
       terminus, all VSGs can be classified into one of three groups: VSGs
       with C-terminal aspartic acid (Class I), VSGs with C-terminal serine
       (Class II), and VSGs with C-terminal asparagine (Class III; 18,25).
(ii)   A hydrophobic C-terminal extension is removed from the C-terminus,
       and the mature VSG is anchored to the lipid bilayer via a covalently
       bound fatty acid (27). This C-terminal anchor consists of a
       dimyristylphosphatidylinositol glycosidically linked to glucosamine
       which is coupled to the polypeptide via additional carbohydrate,
       phosphate, and ethanolamine linkages (28–34). Phosphatidylinositol
       can be released from the VSG by the action of a phospholipase-C
       (35,36). This trypanosome-derived phospholipase-C may thus be
       capable of cleaving the anchor and releasing VSG from the cell
       surface. Phospholipidase-C may be confined to the Golgi complex
       and the base of the flagellum, known as the flagellar pocket (35).
       The flagellar pocket may therefore be part of a VSG membrane
       recycling system in African trypanosomes.
(iii) The 6 Å resolution X-ray structures of the N-terminal two-thirds of
       the VSGs from variants ILTat 1.24 and MiTat 1.2 provide evidence
       that the different VSGs may fold similarly (37). Current models for
       the structure of the VSG coat show the VSGs aligned side by side,
       tightly packed and positioned perpendicularly to the lipid bilayer,
       with the N-terminal portion facing out (38). The elucidation of the
       structure of the VSG which includes a 90 Å long α-helical bundle

will eventually serve as the basis for explaining how the VSG coat can form an impermeable barrier against the host's immune response.

The VSG coat thereby protects the trypanosome in an environment where the presentation of any antigen that is shared between an early and a late trypanosome population would lead to the immediate eradication of the latter.

2.3 The flagellar pocket

The protection provided by the dense VSG coat against immune destruction may impede the free delivery of nutrients to the parasite. In fact, nearly the entire cell surface of the parasite, including its flagellum, is covered with the VSG. However, the flagellar pocket, a unique entity at the base of the flagellum, contains other cell surface proteins in addition to the VSG (39). At this site, a transferrin receptor and a low-density lipoprotein (LDL) receptor were identified (40,41). The flagellar pocket appears to be a small domain within which the uptake of nutrients through receptor-mediated endocytosis, as well as other free communications with plasma, can occur without affecting parasite viability.

3. Ploidy, life-cycle, and chromosome structure

Bloodstream-form trypanosomes are diploid for house-keeping genes and have a DNA content of 0.097 pg per nucleus. Measurements of the complexity indicate that 68% of this DNA is single copy DNA accounting for $2 \times 2.5 \times 10^7$ bp (42–44). The haploid DNA content of the *T.brucei* nucleus is calculated to be 0.041 pg, which is in reasonable agreement with the total amount of 0.097 pg for the diploid *T.brucei* nucleus. The diploid chromosome content of *T.brucei* thus amounts to roughly 70 000 kb of DNA.

Once ingested in a bloodmeal by the insect vector, the parasite differentiates into the procyclic trypanosome which is also diploid (15, 45–47). The procyclic cells then migrate to the salivary glands of the insect vector where they differentiate into metacyclic trypanosomes which can be delivered to a mammalian host with the injection of saliva during feeding (*Figure 3*; 48). Trypanosomes of the species *T.brucei* primarily reproduce through binary fission. Indirect evidence obtained from isozyme analysis suggests that a sexual cycle exists (49,50). The use of DNA probes that detect restriction fragment length polymorphisms has now provided further evidence in favor of genetic exchange (51–56).

Two different models exist which could explain the genetic exchange. In one, the crosses follow Mendelian inheritance, resulting from the fusion of haploid gametes which may be the metacyclic trypanosomes (51). In the second model, the genetic exchange is explained as a result of hybrid

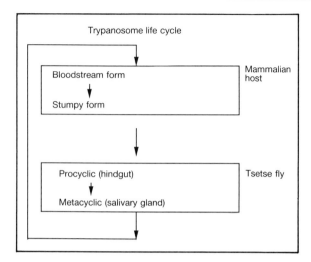

Figure 3.   Schematic representation of the life-cycle of the African trypanosome *Trypanosoma brucei*, which is transmitted by the tsetse fly.

formation due to nuclear fusion in the tsetse fly. The hybrid formation generates a tetraploid cell (52,56). During subsequent divisions, the loss of chromosomes would re-establish a near diploid genome content. It is unclear which of the two processes is responsible for the genetic exchange. All DNA recombinational events that accompany differential VSG gene expression are assumed to be mitotic events.

3.1   Karyotype analysis

Using pulsed-field gradient gel electrophoresis (PFG) to separate chromosome-sized DNA, one can study the organization of the genome of the bloodstream, procyclic, and metacyclic trypanosomes in detail (57 – 62). At first, only about 20% of the diploid genome of *T.brucei* could be separated (58). These molecules consisted of about 100 mini-chromosomes ranging from 50 to 150 kb, five chromosomes of inter-mediate size ranging from 200 to 430 kb, and about five larger chromosomes ranging in size from greater than 630 kb to roughly 1300 kb. The genome of *T.brucei* has now been further separated into a total of 20 bands (*Figure 4*; 63). These bands which range in size from 50 kb to 5.7 Mb (from band 1 with ~ 100 mini-chromosomes to band 19) comprise nearly the total size of the *T.brucei* genome.

Several genes have been mapped to specific chromosomes. The chromosomal locations of active and inactive VSG genes are of particular interest to those studying antigenic variation. The total number of potential (i.e. silent) VSG genes in *T.brucei* may exceed 1000. These silent, or non-transcribed, basic copy (BC) VSG genes are clustered at chromosome internal locations on at least three large chromosomes and

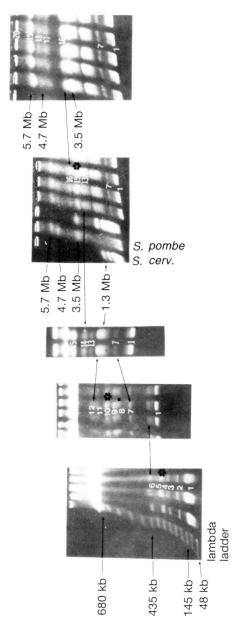

Figure 4. Separation of chromosome-sized DNA from *Trypanosoma brucei* by PFG. The different bands are numbered from the smallest, band 1, with ~100 mini-chromosomes ranging in size from 50 to 150 kb, to band 19, measuring about 5.7 Mb. Bands containing chromosomes with VSG gene expression sites are marked with an asterisk.

are also found at many of the over 200 chromosome ends or telomeres of the mini-chromosomes (58,64 – 68).

The mini-chromosomes (50 – 150 kb) as well as the larger chromosomes appear to be mitotically stable. The mini-chromosomes are confined to the trypanosome species which undergo antigenic variation, and these chromosomes may therefore function solely to expand the repertoire of telomeric VSG genes (58,66,69).

3.2 Karyotype variability

Trypanosome chromosomes show a striking variability in size when a comparison is made between different trypanosome isolates or between variants of the same stock (70,71). This phenomenon is not confined to trypanosomes but also appears in other parasitic protozoa such as *Plasmodium falciparum* (72,73) and *Leishmania* (74 – 77). In *P.falciparum* intrachromosomal rearrangement events as well as the interchromosomal reciprocal exchange of chromosome arms contributes to the variability in the PFG chromosomal separation patterns (78,79). The DNA rearrangement events in trypanosomes are of a diverse nature and will be discussed in detail in the following sections. From the analysis of chromosomal rearrangement events in trypanosomes, it is clear that several of the chromosome bands in PFG gels represent unique copies of chromosomes which are present as single copies in each cell (70). This fact suggests that trypanosomes are aneuploid for several of the smaller chromosomes (200 – 430 kb). The ploidy of the mini-chromosomes is unclear and hence the cell can be considered polyploid for these molecules. The fact that the added numerical values of the chromosome sizes exceeds the total size of the haploid genome content of *T.brucei* suggests that homologous chromosomes in trypanosomes are of different sizes (63).

3.3 Chromosome and nuclear structure

Chromosomes of the parasitic protozoan *T.brucei* are linear molecules. The ends or telomeres of presumably all of the chromosomes share the sequence (GGGTTA)$_n$ (80,81). Telomeres of all eukaryotic chromosomes analyzed to date have variable numbers of such short tandemly repeated sequences (82 – 87). The G-rich strand is always directed 5' to 3' towards the chromosome terminus. The biochemical properties of DNA polymerases lead one to predict that a linear DNA molecule would lose short sequences from its ends at each round of DNA replication. In contrast, however, chromosome ends in trypanosomes and other eukaryotic organisms gradually increase in length (80,88). Blackburn *et al.* (89 – 91) have shown that a similar telomere sequence in *Tetrahymena* DNA serves as a substrate for an enzyme (the telomerase) which adds telomere repeat units to the ends of chromosomes. Thus, a special mechanism is used to maintain telomere length. Presumably a similar system is present in trypanosomes. The telomeric repeats can also undergo drastic

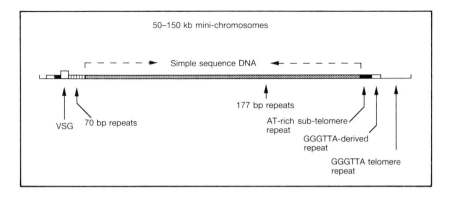

Figure 5. Structure of mini-chromosomes. The main body of the mini-chromosome consists of simple sequence DNA (indicated by the grey box; M.Weiden and L.Van der Ploeg, unpublished data). The GGGTTA telomere repeats (single line), GGGTTA derived subtelomeric repeat (open box) and the AT-rich subtelomere repeat (black box) are schematically represented at both ends of the mini-chromosome. The telomeric VSG gene (open square box) is flanked at its 5′ end by 70 bp repeats (hatched box).

shortening. Heat stress applied to trypanosomes has been implicated as one of the signals that can induce telomere shortening (80,88).

The telomeres in trypanosomes and other eukaryotes are flanked by conserved subtelomeric repeat sequences (80). The telomeres that were analyzed show that the GGGTTA repeats (ranging from several hundreds of base pairs to several kilobase pairs) are flanked by a diverged GGGTTA-derived repeat that is located next to an AT-rich stretch of DNA (several kilobases long).

The analysis of the smallest *T.brucei* mini-chromosomes and the telomere analysis have allowed the determination of a structural model for the mini-chromosomes (*Figure 5*; 58,70,80, and M.Weiden and L.Van der Ploeg, unpublished data). The internal sequence of the mini-chromosome probably consists of a 177 bp satellite repeat sequence (58,92,93). This repeat sequence may function to generate mini-chromosomes that are of sufficient length to render the molecules mitotically stable. Mini-chromosomes thus have a limited protein coding potential and mainly consist of simple sequence DNA. Centromeres have not yet been identified. Mini-chromosomes encode an abundance of VSG genes which are located on their telomeres and the different VSG genes are flanked by 70 bp repeats at their 5′ end.

From the analysis of trypanosomes that had been passed through the tsetse fly and had undergone a sexual cycle, it is clear that the mini-chromosomes do not follow a Mendelian pattern of inheritance (52). The repertoire of mini-chromosomes can be rapidly expanded following a genetic cross. Evidence for this is provided by the presence of trypanosomes which contain roughly twice the amount of mini-chromosomal DNA. Whether this observation reflects the fact that these chromosomes do not always segregate properly at meiosis (or mitosis) or that this repertoire of molecules can be amplified is still unclear.

The structures of the larger chromosomes have not been analyzed in detail. Two features are worth noting.

(i)   The rDNA genes are distributed over at least six different chromosomes indicating that the nucleolus may gather rDNA repeats from different chromosomes.

(ii)   The house-keeping genes identified so far are confined to the larger chromosomes of *T.brucei* (63).

The higher order organization of chromosomes and their mode of segregation during mitosis in trypanosomatidae was revealed by three-dimensional reconstitution of sectioned nuclei, which were analyzed by electron microscopy (94–96). During mitosis in the protozoan *T.cruzi*, the chromosomes are distributed into ten electron-dense, plaque-shaped structures. These electron-dense, chromatin-containing plaques divide into half-plaques during metaphase and move to the poles during anaphase. The plaques are attached to two sets of approximately 60 microtubules. The chromosomes in *T.brucei* comprise about 100 mini-chromosomes and at least 18 larger chromosomes (63). These chromosomes appear to be distributed over plaques attached to roughly 30 microtubules (96). The number of chromosomes thus appears to exceed the number of plaques and microtubules. The migration of half-plaques at mitosis therefore indicates that multiple chromosomes are compacted into a single plaque, giving rise to the ordered segregation of the chromosomes.

4. DNA recombinational mechanisms in differential VSG gene expression

Of paramount importance to the success of antigenic variation is the fact that the VSG coat contains all the antigenic determinants exposed on the live parasite. The expression of a single new VSG gene can thus alter the entire antigenic composition of the cell surface coat. The switch from one VSG coat to the next is a process which is believed to occur independently of external stimuli (i.e. the immune response) and in *T.brucei* has been shown to occur at a frequency of about 10^{-6} to 10^{-7} per cell division (10,16,97,98). Unfortunately, this switch frequency is too low to allow a direct analysis of the process of switching. For this reason all studies of the mechanisms of antigenic variation have been performed by comparing the transcribed VSG gene of a parent population to the transcribed VSG gene of its daughter population(s).

The process of antigenic variation is controlled at the transcriptional level through the sequential activation of different VSG genes. Each trypanosome usually expresses only a single VSG gene at any given time,

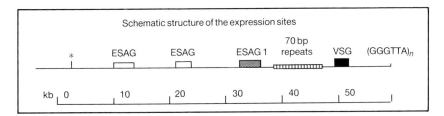

Figure 6. Schematic representation of the structure of an expresssion site (ES). The telomeric expression site transcription unit with the VSG gene is large and measures over 50 kb. Several ESAGs exist in addition to the VSG gene. The asterisk marks the 5′ border of the expression site transcription unit.

explaining the efficient use of the repertoire of VSG genes throughout an infection. In addition, the mutually exclusive activation of VSG genes prevents the situation in which antibodies directed against previously expressed VSG coats kill trypanosomes with a mixed coat.

4.1 Structure of expression sites

An important characteristic of the VSG gene being transcribed is that it is invariably located at a telomere (99). When comparing the chromosomal location of transcribed telomeric VSG genes, it becomes clear that these genes are confined exclusively to the telomeres of a few specific chromosomes (58,63,70). Structural analyses of these telomeres show that each of them actually contains several tandemly linked genes of which the active VSG gene is the most distally located (100 – 107). The expression of the telomeric VSG gene and the expression site-associated genes (ESAGs) is coordinately regulated at the transcriptional level (100 – 103). These telomeres thus represent conserved VSG gene expression sites of which only one is transcriptionally active at any given time (*Figure 6*). The ESAGs, however, differ from the VSG genes in that they do not translocate in and out of expression sites (see following sections).

The first ESAG identified (ESAG1) is located at the expression site on a 1.5 Mb chromosome and directly upstream of a large region with tandem arrays of a 70 bp repeat (100). ESAG1 is transcribed into a 1400 nt mRNA and encodes a membrane protein of undetermined function (108). Subsequently, several genes similar to ESAG1 have been found in other expression sites. In one expression site, the ESAG1 reading frame is interrupted by frame-shifts leading to premature translational termination (109). Since approximately 15 copies of ESAG1 have been found, up to 15 expression sites may exist (100). These ESAGs are thus invariably and constitutively expressed genes of particular expression sites.

Kooter *et al.* (101) analyzed the expression site containing the VSG 221 gene located on a 3 Mb chromosome. They characterized about 55 kb of DNA upstream of the VSG 221 gene and identified several other ESAGs in this locus. A third expression site was analyzed with VSG gene 1.8

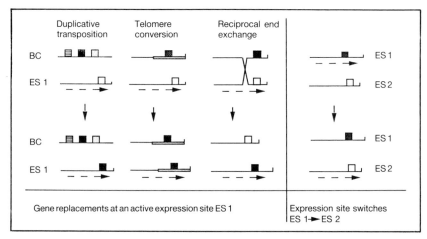

Figure 7. Schematic representation of the DNA recombinational events associated with antigenic switches. (Left) DNA recombinational events in an already active expression site. (Right) Antigenic switches that result from the inactivation of one (ES 1) and the activation of a new expression site (ES 2). Square boxes represent VSG gene coding sequences. The dashed lines and arrows indicate the transcription at the telomeric expression site.

located on a 430 kb chromosome (102). At least three potential ESAGs were isolated from a 430 kb chromosome-specific library. Finally, Pays and co-workers have studied the expression sites containing the VSG gene of variant AnTat 1.3 located on a 230 kb chromosome (103). The overall structure of these expression sites resembles that of the previously studied variants with at least three ESAGs located upstream of the VSG gene.

4.2 General description of antigenic switches

VSG genes which are neither transcribed (BC VSG genes) nor located at a telomere can become activated by acquiring a position in an active expression site through different modes of genetic recombination. Obviously, if there were only one VSG gene expression site, the differential activation of the VSG genes could be fully explained by gene replacements at this unique expression site. In that case, the loss or displacement of a previously transcribed VSG gene would assure the transcription of only a single VSG gene at any given time. However, multiple expression sites located on different chromosomes were identified in various *T.brucei* strains, and it is their transcriptional activity which is controlled in a mutually exclusive manner. New VSG genes can thus become activated through either the activation of a new expression site already containing a VSG gene, or through the recombinational events which place a new VSG gene in an expression site which is already active (*Figure 7*).

A specific repertoire of VSG genes is displayed early on during an infection. These comprise VSG genes that are already activated in the metacyclic trypanosomes present in the salivary glands of the tsetse fly (110 – 112). During a chronic infection, new trypanosome variants also arise in a somewhat pre-determined order made up of a preferential early and a late population of antigen variants (113,114). The observation that the sequential expression of VSG genes occurs in an order which is not perfectly repeated and not absolutely predictable is poorly understood (115 – 118). This order can be affected by the species of the host in which the parasites proliferate (118). An immunological selection may therefore play a role in determining the order in which particular variants arise during an infection.

4.3 Antigenic switches as a result of VSG gene displacement at an active expression site

DNA recombinational events can result in the translocation of a BC VSG gene into an active expression site. The VSG gene copy that is generated and transcribed is called the expression-linked copy (ELC; 119 – 122). In this type of antigenic switch the same expression site retains its transcriptional activity and now expresses a new VSG gene (58). Translocation of a BC VSG gene into an active expression site can occur via three mechanisms:

(i)   duplicative VSG gene transposition (119,121 – 127);
(ii)   telomere conversion (128);
(iii) the reciprocal exchange of chromosome arms containing VSG genes (129).

The mechanisms of these DNA recombinational events which replace the previously expressed VSG gene with a new VSG gene copy will be discussed first. The second type of regulatory event whereby an antigenic switch can be carried out by the activation of a 'new expression site' and the inactivation of the 'old expression site' will be analyzed later.

4.3.1 VSG gene activation by duplicative transposition; structure of the transposed segment

Basic copy VSG genes located at internal positions within the chromosome or at telomeres can be duplicatively transposed to a VSG gene expression site thus generating an ELC (*Figure 8*). The previously active VSG gene residing at the telomeric expression site is lost in the process and is presumably destroyed. This process ensures the transcription of a single VSG gene at any given time (127,130,131).

The structure of the transposed segments suggests potential mechanisms of VSG gene transposition. The transposed segments of a single VSG gene can be of variable size (132 – 139). This size variation results from the fact that the transposed segments contain the VSG gene coding sequence of about 1500 bp as well as a variable stretch of DNA, ranging from several hundred base-pairs to several kilobase pairs located upstream of the VSG gene (*Figure 8*). The 5' boundary of the transposed

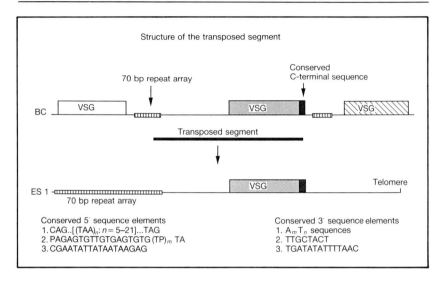

Figure 8.   Basic copy VSG genes arranged in tandem arrays can be duplicatively transposed to a telomerically located expression site. The transposed segment is flanked by conserved sequences, a 5′ 70 bp repeat (its consensus sequence is indicated at bottom left) and 3′ located conserved C-terminal sequences and untranslated extension sequences.

segments is thus located in divergent DNA sequences. However, common target and recipient DNA sequences do occur. A DNA sequence frequently found flanking the 5′ end of many transposed segments is an imperfect 70 bp repeat (140–143). This repeat sequence flanks most BC VSG genes, whether they are located at positions within the chromosome or at telomeres. Both types of VSG genes can be activated by duplicative transposition (144,145). The 70 bp repeat sequence is also present in front of the ELC at the expression site. Here it consists of much larger arrays of 70 bp repeats (up to 25 kb) (121,141,146). Therefore the 5′ end of the transposed segment shares homology with the target (ELC) sequences. The 70 bp repeat sequence may serve in aligning the incoming gene with the expression site. It has also been suggested that the 70 bp repeat sequence may adopt non-B-DNA conformations, which could play a role in initiating the recombinational process (*Figure 8*; 140).

   Other non-70 bp repeat sequences at the 5′ ends of transposed segments may be involved in mediating the duplicative transposition as well. Nucleotide sequence homology is obtained in these cases because novel DNA sequences, having homology to the 5′ end of the transposed segment, had previously been duplicatively translocated into the expression site (*Figure 9*). Thus extensive sequence homology (hundreds of base pairs) still exists between the donor (BC) and proposed target expression site sequences (133,139). It can be assumed that all these DNA sequences serve to align the incoming gene with the ELC.

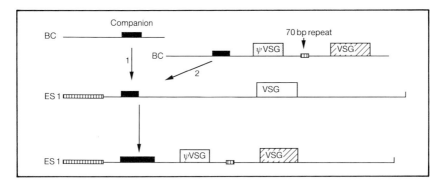

Figure 9. Sequences other than 70 bp repeats which flank many BC VSG genes can make up the 5' end of the transposed segment. These DNA sequences were first transposed into an expression site (transposition event 1). Presumably, the DNA sequence homology between these telomeric companion sequences, and the related DNA sequences located 5' of the VSG gene can now serve to mediate duplicative transposition into the telomeric expression site (transposition event 2). The figure displays transposition of a VSG pseudogene and a VSG gene on a single large transposed segment (133).

Homologous DNA sequences at the donor and target sites exist at the 3' ends of the transposed segment as well (*Figure 8*; 147 – 149). The nucleic acid sequence homology results from the fact that all VSG genes share extensive sequence homology in the last 50 amino acids of the coding sequence (24 – 26). Homology persists throughout the 3' untranslated extension of VSG genes where fully conserved 8-mer and 14-mer sequences are found (130 – 132). The 3' ends of the transposed segments can therefore have a variable location, either within the 3' end of the coding sequence of the VSG gene or beyond the 3' end of this coding region (120). It is therefore assumed that the new 3' end found in some VSG gene ELCs is a remnant of the VSG gene that previously occupied the expression site.

The repetitive elements at the 5' ends and the repeated sequences at the 3' ends of the transposed segments do not share sequence homology. Therefore, the structure of these transposed segments does not resemble those of transposons or retroposons which are characterized by inverted repeats or direct repeats and a poly[A]$^+$ tail at one of their borders (150,151). In fact, the structure of these transposed segments resembles that of the duplicatively transposed segments in the yeast mating type locus (152 – 156). In this locus, the a or α mating type genes (HML and HMR) are translocated like cassettes into a single mating type locus (MAT). Genetic information at MAT is thus changed from a to α or α to a. These duplicative transpositions in yeast are the result of a process of unidirectional gene conversion. This mechanism for the generation of genetic diversity is also found in other eukaryotes. For example the light chain variable immunoglobulin genes in chickens are arranged in tandem arrays of pseudogenes located upstream of the functionally rearranged and transcribed immunoglobulin light chain gene (156). A functional and diverse repertoire of light chains is generated through unidirectional gene conversion of the pseudogene sequences into the last variable light chain gene of the array.

On the basis of the structures of the transposed segment and the expression site, it can be assumed that duplicative transpositions of VSG genes occur by unidirectional gene conversion. Unlike the yeast mating type switch and the generation of the chicken immunoglobulins, DNA intermediates of the switching process have not been identified in trypanosomes. This is presumably due to the less frequent occurrence of duplicative transpositions in trypanosomes (<1 per 10^6 cell divisions).

Aside from the common structures of the loci where duplicative transpositions occur, several important differences exist between VSG gene transposition and other types of transpositions.

(i)   The yeast mating type and chicken immunoglobulin gene rearrangements are intrachromosomal rearrangement events. In contrast, the duplicative transpositions of VSG genes can occur as interchromosomal events, presumably during mitosis (58,63). For instance, the duplicative transposition of the BC VSG 118 gene in variant 118a takes place between a 3 Mb chromosome containing the BC and a 1.5 Mb chromosome containing the expression site. Duplicative VSG gene transpositions are best described as ectopic (between different chromosomes) and homologous (between related but not identical DNA sequences) gene conversion events (Bailis and Rothstein, personal communication).

(ii) In the yeast mating type switch, the translocated genes are transcribed due to the activation of a co-transposed promoter. Transcriptional activation of VSG genes by duplicative transposition may result from the activation of a co-transposed promoter or from the juxtaposition of a transposed gene with a promoter that is already located in the expression site far upstream of the transposed segment (157, 158).

4.3.2 VSG gene activation by telomere conversion

A second type of DNA recombinational event exists by which VSG genes can be translocated into an already active expression site. This type of DNA recombinational event involves BC VSG genes which already reside at telomeres. In this event, the transposed segment seems to carry the VSG gene and all of the flanking telomere repeat sequences (128). The conversion event thus differs from duplicative transposition in that it extends towards the tip of the chromosome and does *not* resolve in the vicinity of the 3' end of the translocated VSG gene (*Figure 10*). The 5' boundary of these telomere conversion events is presumably located in the imperfect 70 bp repeats that flank telomeric VSG genes. Telomere conversion, therefore, resembles duplicative unidirectional gene

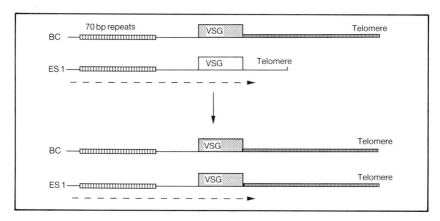

Figure 10. Telomerically located BC VSG genes can enter an active expression site by telomere conversion and thus become transcribed. The recombinational event is presumably initiated in the 70 bp repeats and extended towards the tip of the telomere. The dashed arrow indicates transcription at the active expression site.

conversion with respect to the 5' boundaries, but involves larger stretches of DNA at the 3' end.

4.3.3 VSG gene activation by reciprocal exchange

The last example of activation of a telomeric BC VSG in an already active expression site is the reciprocal exchange of chromosome arms. This switch event differs from the previously described gene conversion events since it does not involve an alteration in VSG gene copy number. The reciprocal recombination merely involves an exchange of telomeric VSG genes. Only one example of a reciprocal exchange was identified in the antigenic switch from variant AnTat 1.3A to AnTat 1.10D (*Figure 11*; 129). The previously active VSG gene 1.3A of the expression site was displaced to a new chromosome while the 1.10 telomere was placed in the expression site and encoded a new VSG coat.

Apparently, any of these three different recombinational mechanisms can be used to replace a VSG gene at an already active expression site. These DNA recombinational events do not seem to affect the transcriptionally active status of the expression site. The VSG gene transpositions are simply targeted to a chromosome containing a transcriptionally active expression site.

4.4 Antigenic switches resulting from differential transcriptional control of expression sites

VSG gene activation events which have already been discussed result from the translocation of a previously silent (BC) VSG gene into an already transcriptionally engaged VSG gene expression site. If there were only one VSG gene expression site in the genome, the mutually exclusive trans-

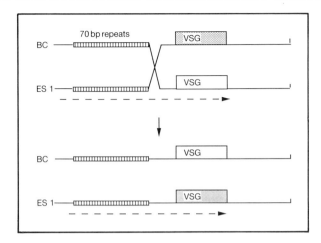

Figure 11.   Telomerically located BC VSG genes can enter an active expression site by the reciprocal exchange of chromosome arms. The dashed line and arrow indicate transcription at the active expression site.

criptional control of VSG genes could be easily explained by these gene displacement events. However, multiple independent expression sites exist which are located on chromosomes of drastically different sizes (58, 63,66,159 – 166). Three of these expression sites have thus far been identified in *T.brucei* MiTat stock 427-60 on chromosomes of 430 kb, 1.5 Mb, and 3 Mb, respectively (see *Figure 4*, bands marked with asterisks; 63). The transcription of the expression sites must, therefore, be controlled in a mutually exclusive manner.

4.4.1 Mutually exclusive transcriptional control of expression sites

Antigenic switches that result from the activation of a new expression site and the inactivation of the old expression site require coordinate control of these two events. This becomes obvious when one considers the fact that 'naked' cells, which do not transcribe any VSG gene expression sites, will be lysed by the activation of the alternate complement pathway (18). In addition, cells with two active expression sites are also potentially at risk due to antibodies directed against the 'old' coat that will lead to cell death even though the 'new' coat is stably expressed.

Several different events have been identified that shed light on the mechanisms involved in the differential transcriptional control of expression sites. Expression site activation has been observed as either:

(i)   an *in situ* activation (159 – 166); or
(ii)   the activation of an expression site associated with its translocation to a new chromosome (70).

A comparable repertoire of events has been observed during the inactivation of expression sites which result from:

(i)   an *in situ* inactivation of an expression site (167 – 170);
(ii)   the destruction of the expression site due to deleterious mutations which eliminate transcription of the VSG gene without affecting transcription of the ESAGs (171); and
(iii) the inactivation of an expression site associated with the reciprocal exchange of chromosome arms thus moving the site to a different and a larger chromosome (70,172).

These events will be briefly illustrated with specific examples and the mechanisms that may be involved in their control will then be discussed.

4.4.2 In situ activation of expression sites

Several antigenic switches have been identified in which the activation of a VSG gene does not result from the duplicative transposition of a VSG gene with the generation of an ELC. The VSG genes that are activated in this manner are shown to be already located at telomeres and are apparently activated *in situ* (159 – 166). Another example of a coupled *in situ* activation and inactivation was identified in the study of VSG genes that had previously been activated by duplicative transposition of a VSG gene into an expression site (167 – 170). These VSG genes are most often lost during the next antigenic switch due to their displacement by the new VSG gene (as described in Section 4.3.1). However, in some antigenic switches, this VSG gene and its expression site are simply retained at the same chromosomal location but become inactive (*Figure 12*). A new expression site, located on a different chromosome, is activated and this site now takes over the production of the VSG coat. These data indicate that the mutually exclusive transcriptional control of VSG genes may involve the coordinate transcriptional regulation of different expression sites. However, there are exceptions to this rule given the existence of trypanosomes with two simultaneously active expression sites (see Section 4.4.5).

A search for cryptic DNA rearrangement events that may explain transcriptional control of expression sites has been unsuccessful. Long-range restriction enzyme maps of these expression sites also failed to reveal small DNA rearrangement events indicative of a movable promoter or movable transcriptional control element. However, other structural changes can be observed. The active expression site, in contrast to the inactive site, can be shown to lack secondary base modifications (173, 174). The exact nature of these modifications remains obscure. Another structural feature which differs between transcriptionally active and inactive sites is the length of the telomeric repeats. The telomeric repeats measure several kilobases in an active expression site whilst in inactive expression sites they measure up to tens of kilobases (175,176).

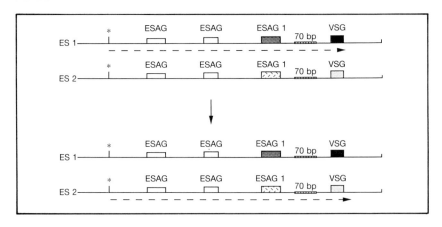

Figure 12. Antigenic switches without apparent DNA recombinational events can result from the *in situ* activation and inactivation of expression sites located on different chromosomes. The dashed line and arrow indicate transcription at the active expression site.

4.4.3 Translocation of an expression site

From the previous summary, it becomes obvious that expression sites located on different chromosomes can be transcriptionally controlled in an apparently coordinated manner. However, in addition to the coordinately controlled *in situ* activation and inactivation of an expression site, another expression site switch exists in which the transcriptional activation appears to be controlled by a DNA translocation event (*Figure 13*).

In this switch, the activation of the new expression site occurs concomitantly with its duplicative translocation to a small 340 kb

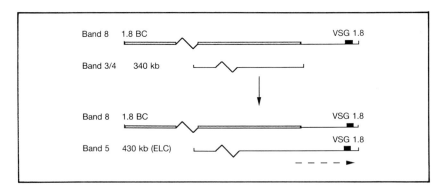

Figure 13. Activation of a VSG gene expression site sometimes correlates with the translocation of the site to a different chromosome. The numbers of bands refer to the location of bands in the chromosome separation pattern in *Figure 4*. The black box represents the 1.8 VSG gene. The dashed line and arrow indicate transcription at the active expression site.

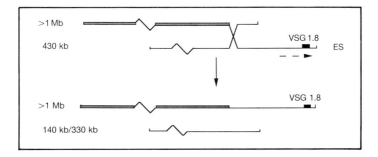

Figure 14. Inactivation of an expression site sometimes correlates with the translocation of the site to a different chromosome. Abbreviations are as indicated in *Figure 13*.

chromosome (70). The stretch of DNA that is duplicatively transposed is large and measures about 90 kb. The mechanism by which these DNA recombinational events might control the transcription of expression sites is unclear. Since active expression sites lack secondary base modifications and have short telomeres, it is possible that chromosomal rearrangements affect transcription through positional control.

4.4.4 Reciprocal exchange of chromosome arms leading to expression site inactivation

An additional type of recombinational event has been described involving the inactivation of the VSG gene 1.8 expression site located on a 430 kb chromosome. The inactivation of this expression site occurs simultaneously with its translocation to a new chromosome via reciprocal end exchange of the site (172). Two independent translocations were studied involving the movement of stretches of DNA measuring 80 and 240 kb (*Figure 14*). This type of DNA recombinational event is in agreement with the model stating that DNA recombinational events can affect transcription at expression sites through positional control.

4.4.5 Mutations that disable the expression site and potential switch intermediates

Antigenic switches that occur as a result of the activation of one expression site and the inactivation of another should have potential switch intermediates with two active expression sites. Only a few examples of potential switch intermediates exist (171,177,178). These examples are informative and show that expression site activation and inactivation are not necessarily *directly* coupled. With the use of an immunological selection against the 221 VSG coat, a trypanosome was isolated which expressed a new VSG on the cell surface. However, the 221 expression site was not inactivated (171). The 221 VSG gene expression site was retained in a transcriptionally *active* form but contained a large insertion of DNA

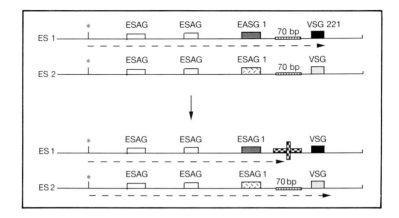

Figure 15. Two expression sites can be active at the same time. In one case studied, the antigenic switch proceeded successfully because the transcription of the VSG 221 gene was prevented, presumably due to the presence of a large insert of DNA in front of the VSG 221 gene (indicated by the hatched cross). A different expression site had taken over the production of the VSG coat.

located just upstream of the 221 VSG gene (*Figure 15*). This insertion prevented the 221 VSG gene from being transcribed while the ESAGs, located upstream of the insertion, were still transcribed normally. A different expression site had taken over the production of the new VSG coat. Apparently, selection against the presence of the 221 VSG coat resulted in the isolation of a trypanosome clone in which a mutation simply prevented the production of 221 VSG.

From the functional inactivation of this VSG 221 expression site located on a 3 Mb chromosome, two conclusions can be drawn.

(i)   Transcription of expression sites is not always mutually exclusive and more than one expression site can be transcribed at the same time; and

(ii) activation and inactivation of expression sites can occur as independent events.

The current understanding of the transcriptional control of expression sites supports a model in which *in situ* control of the expression sites is the regular route of transcriptional regulation. However, DNA recombinational events at expression sites may occasionally be utilized by the organisms, since they can facilitate the expression site switch process in case the site was not inactivated (or activated) through the 'regular' transcriptionally controlled route (see Section 6). This is most easily visualized if one considers the fact that the continued expression of a VSG gene against which antibodies are directed is potentially lethal to the cell. Antibodies may thus select for cells in which the expression of a particular VSG gene was prevented by any of a variety of different DNA recombinational events.

4.5 Control of expression site switching

Antigenic switches resulting from the duplicative transposition of a VSG gene into an already active expression site are explained relatively simply by the concept of gene replacement. Antigenic switches resulting from the differential transcription of expression sites are more complicated since they require the sequential activation and inactivation of separate expression sites. Potential switch intermediates with two active expression sites and two VSGs in the coat have occasionally been found. This observation led to the hypothesis that expression sites may be switched on and off independently.

In formulating a model which explains coordinate control of VSG gene expression sites and independent control of expression site switches, an immunological selection against switch intermediates has been proposed (133). Such selection could act on two different types of switch intermediates.

The first type of intermediate consists of switches that result from gene replacement in an already active expression site. These switches should have a *short-lived* VSG double coat containing the old and the new VSGs, while the new VSG replaces the old coat.

The second type of intermediate consists of switches that result from the activation of a new expression site and the subsequent inactivation of the old expression site. These intermediates will contain two simultaneously active VSG genes and may therefore be *long lived*. Such trypanosomes which stably express two VSG genes have indeed been identified (175,176). The life-span of these 'double expressors' and their frequency of occurrence in the population of trypanosomes must depend on the probabilities with which expression sites are turned on and off. It is obvious that the frequency with which any expression site is turned on is expected to be higher than the frequency with which the expression site is turned off. In this way, the generation of naked trypanosomes that are non-viable and would be immediately eliminated can be minimized.

Immunological selection is proposed to act similarly on both types of doubly expressing trypanosomes. However, as discussed above, the efficiency of selection may be affected by the period of time for which the intermediates exists. Furthermore, selection could act on trypanosomes which are in the process of switching by killing the cells that still have the 'old' coat. Selection may additionally act on trypanosomes with a double coat by killing cells that express any combination of VSGs that result in a non-protective double coat (a result of structural gaps leading to exposure of common antigenic determinants or activation of the alternate complement pathway). Terminating transcription quickly at the 'old' expression site is an obvious selective advantage. One can thus explain how the different types of DNA rearrangement events leading to the inactivation of the expression site are selected, since they allow survival of the cells.

An immunological selection against switch intermediates may thus affect the trypanosome populations that arise during a chronic infection. In addition, growth rate competition among different variants may play a role in the outgrowth of new variants. Mathematical modelling of parasitemias, in which growth rate competition is chosen as the single parameter, does not support the hypothesis that growth rate competition is the sole parameter determining the growth profile of the observed parasitemias. However, a model which adds selection against switch intermediates does mimic the situation *in vivo*, producing waves of relapsing parasitemias. (179,180).

5. Identification of transcriptional initiation sites

The identification of VSG gene promoters is fundamental to understanding the control of expression site transcription. However, the transcriptional control of protein-coding genes, including the VSG genes, in African trypanosomes and related species is still poorly understood. This is mainly due to the fact that the mechanisms of transcription of protein-coding genes and maturation of mRNAs in these organisms are dramatically different from those in other eukaryotes. In addition, progress in the study of antigenic variation has been hindered by the slow progress in establishing a reliable DNA transfection system (181,182,260).

5.1 Discontinuous transcription and *trans*-splicing

In order to understand VSG gene transcriptional control it is essential to consider the peculiar routes of pre-mRNA maturation in trypanosomes. In trypanosomes and related Kinetoplastida, every mRNA consists of two exons (183,184) which are initially transcribed from separate genes and are usually located on different chromosomes (69,185). The first exon is a capped (185 – 189) 5' non-coding exon (the mini-exon or spliced leader) which is common to *all* mRNAs (190 – 193). This mini-exon is derived from the first 39 nt of a 140 nt mini-exon donor RNA (medRNA) transcript encoded by approximately 200 copies of identical repetitive genes (194 – 200). An abundance of experimental data outlined below proves that the joining of the mini-exon and the coding exon occurs through the *trans*-splicing of the independently transcribed precursor RNAs (*Figure 16*).

(i)   The presence of the canonical splice-donor and splice-acceptor AG – GT sequences at the exon – intron and intron – exon junctions, respectively, implies that the exons are indeed joined by RNA splicing (for reviews see refs 7 and 8). The 3' splice-acceptor site is characterized by the presence of multiple, closely linked, and alternative, splice-acceptor sites, as well as potential NYY(R)AY branchpoint sequences and a polypyrimidine stretch at approximately – 60 nt upstream of the 3' splice site (201).

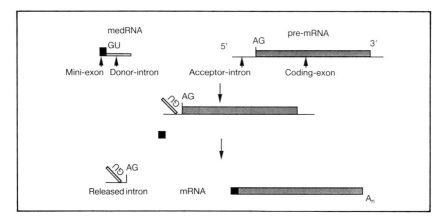

Figure 16. Schematic representation of the process of *trans*-splicing. Mini-exon and coding exon joining in Kinetoplastida is mediated by RNA splicing of two separately transcribed RNA precursors (medRNA on the left and pre-mRNA with the coding exon on the right). Cleavage at the 5′ splice site of the medRNA generates a 39 nt 5′ capped mini-exon (black box) and releases the 105 nt 5′ donor-intron. Branched Y-shaped intron intermediates and side products of the splicing reaction are generated, instead of the regular lariat-shaped intron that results from *cis*-splicing. The medRNA presumably performs both a catalytic and a substrate function in the *trans*-splicing reaction. GU and AG represent the first two nucleotides of the medRNA derived 5′ intron and the last two nucleotides of the 3′ acceptor intron, respectively.

(ii)   The medRNA genes and several protein-coding genes are transcribed discontinuously, as shown by the fact that several protein-coding genes are located on chromosomes without medRNA genes. Therefore, precursor RNAs from different chromosomes are joined post-transcriptionally with the 39 nt mini-exon (70,185).

(iii)   Some protein-coding genes are transcribed by α-amanitin-insensitive RNA polymerases, while the mini-exon is transcribed by an α-amanitin-sensitive RNA polymerase which is presumably an RNA polymerase II. Thus different RNA polymerases may produce the two precursor RNAs (202).

(iv)   Measurement of the half-life of putative high molecular weight splice intermediates gives values that appear too short to support an alternative model of exon joining, that is priming ligation (203).

(v)   Analysis of branched polyadenylated pre-mRNAs and of released intron side-products provides evidence in favor of *trans*-splicing in a process that is mechanistically similar to *cis*-splicing. These molecules were predicted to have Y-shaped structures, analogous to the lariat-shaped introns of *cis*-splicing. Y-shaped pre-mRNAs and intron side-products of the splicing reaction have indeed been identified (204–207). All genes in trypanosomes, including the VSG genes, lack regular introns that would be removed by splicing in *cis*.

5.1.1 Trans-splicing and snRNAs

In partial fulfilment of the requirements of the enzymatic machinery involved in *cis*-splicing in higher eukaryotes, small nuclear RNAs (snRNAs) involved in splicing (snRNAs U2, U4, and U5; 208,261) have been identified in trypanosomes. In addition, it was also possible to immunoprecipitate the medRNAs from various organisms in which pre-mRNAs mature through *trans*-splicing using antibodies directed against Sm antigen associated with small nuclear ribonucleoprotein particles (snRNPs; 209–211). This observation led to the proposal that the medRNA is in fact an snRNA. The medRNA was subsequently thought to be the analog of the U1 RNA involved in *cis*-splicing. This belief is supported by the finding that a consensus sequence for an Sm antigen-binding site was found in several of the medRNAs of these protozoa and in the medRNA of the nematode *Caenorhabditis elegans* (210–212). The medRNA may also fold in a manner similar to that of U1 RNAs of other eukaryotes (212). In addition, the *trans*-spliced leader in *C.elegans* has the trimethyl (M3 ^{2,2,7}G) cap, characteristic of most snRNAs. Both the methyl-G cap of the medRNA as well as the cap in mature mRNAs in trypanosomes have a different and peculiar structure: they are flanked by five additionally modified bases (186–189). The function of this unusual modified cap in trypanosomes ($M^7GpppA*A*C*U*AA*CG$; asterisks denote modifications; 186) is unclear. The medRNA of *C.elegans* as well as the medRNA of *T.brucei* and related protozoa thus appear to be analogous to the U1 snRNAs involved in *cis*-splicing in other eukaryotes. The medRNA itself is expected to assemble into the spliceosome as its U1 component, presumably performing both a catalytic and substrate function.

5.1.2 Polygenic pre-mRNA and trans-splicing

Maturation of mRNAs through *trans*-splicing allows the addition of a 5' cap to precursor RNAs. Many protein-coding genes in trypanosomes have now been shown to be arranged in tandem arrays (VSG, heat shock protein 70 kd, calmodulin, phosphoglycerate kinase, and α- and β-tubulin genes; 201,213–220) and in the cases of several of these arrays, it could be shown that they are initially transcribed as larger pre-mRNAs containing multiple coding exons (*Figure 17*; 158,213,214,218–220). These polygenic pre-mRNAs are believed to be the precursors of the mature mRNAs. Maturation of larger pre-mRNAs through *trans*-splicing with medRNA may thus lead to the generation of multiple mature mRNAs from a single precursor RNA. It is still unclear where the transcription of these polygenic transcription units initiates.

The maturation of polygenic pre-mRNA requires cleavage for poly[A]$^+$ addition and *trans*-splicing in the short intergenic regions. Analysis of the trypanosome mRNAs has already shown that the eukaryotic AAUAAA polyadenylation signal used by higher eukaryotes is not found at the end

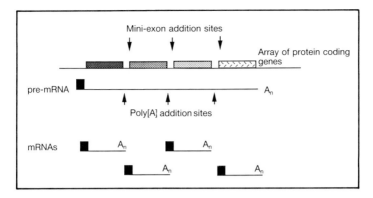

Figure 17. *Trans*-splicing of the medRNA may generate multiple mature mRNAs from a single pre-mRNA. It is unclear whether the splicing event occurs at the nascent RNA level, or whether mini-exon-containing, partially matured pre-mRNAs (as indicated in the figure) can serve as a substrate of the *trans*-splicing reaction. Abbreviations are as in *Figure 16*.

of trypanosome mRNAs (24 – 26,215). The mechanisms and timing of the different processing events in the intergenic regions are therefore still unclear.

All attempts to develop splicing and transcription systems to study these processing events *in vitro* have been unsuccessful. These efforts have, however, led to the identification of several novel enzymatic activities involved in the efficient 3' RNA end-labeling, RNA ligation, and exonucleolytic degradation of RNA other than mRNA (221 – 223). The relevance of these findings for RNA metabolism in trypanosomes is still obscure.

5.2 Polygenic transcription of VSG gene expression sites by α-amanitin-insensitive RNA polymerases

As in mRNA of other protein-coding genes analyzed in trypanosomes, VSG mRNAs consist of two separate exons initially transcribed from the medRNA gene and the VSG gene which are located on different chromosomes (70,185). The 5' end of the VSG mRNA therefore does not identify the transcription initiation site of the VSG gene but that of the medRNA genes.

Adding to these difficulties is the possibility that the transcription of VSG gene expression sites is regulated in a manner that differs drastically from the transcriptional control of other protein-coding genes in trypanosomes. Unlike the majority of protein-coding genes in trypanosomes and other eukaryotes, the expression sites are transcribed by an RNA polymerase that is insensitive to the drug α-amanitin at a concentration up to 1 mg ml^{-1} (202). This level of drug resistance is also displayed in the transcription of rRNA genes by RNA polymerase I in trypanosomes and other eukaryotes.

5.2.1 RNA polymerases in trypanosomes

On the basis of the sensitivity of transcription to the drug α-amanitin, three different RNA polymerases have been identified in eukaryotic cells (224):

(i)   RNA polymerase I which transcribes rRNA genes and is resistant to concentrations of α-amanitin of 1 mg ml^{-1};

(ii) RNA polymerase II which transcribes protein-coding genes and several snRNA genes. Transcription by RNA polymerase II is inhibited by α-amanitin at concentrations as low as 5 μg ml^{-1};

(iii) RNA polymerase III which transcribes tRNA genes and which is sensitive to intermediate levels (500 μg ml^{-1}) of α-amanitin.

In trypanosomes, three such RNA polymerases are found (225 – 230,262).

The large subunit for each of the three types of RNA polymerase genes were identified in the genome of *T.brucei* using the conserved nucleotide sequences of other eukaryotic RNA polymerase genes as probes. Two polymorphic loci with different genes encoding the large subunit for RNA polymerase II (229,262) and an RNA polymerase III gene (230) have since been characterized at the nucleotide level. The two types of RNA polymerase II genes differ by three amino acid substitutions involving residues 131 (valine to methione), 391 (valine to isoleucine) and 472 (serine to asparagine). A characteristic feature of the C-terminal sequence of RNA polymerase II in yeast, *Drosophila*, mouse, and hamster is the presence of several tandemly repeated amino acid sequences. In yeast, 26 repeats with the consensus Tyr-Ser-Pro-Thr-Ser-Pro-Ser are present. The trypanosome RNA polymerase II sequence shares no structural homology with these repeats aside from the presence of frequent proline, serine, and tyrosine residues. In addition, the trypanosome C-terminal sequence has a high frequency of acidic amino acids.

The three types of trypanosome RNA polymerases have been separated using column chromatographic techniques (227 – 229; Cornelissen *et al.*, personal communication). In this manner, a single type of RNA polymerase (RNA polymerase I) has been fractionated which is resistant to an α-amanitin concentration of 1 mg ml^{-1}.

5.2.2 Transcription of VSG genes

Transcription of all protein-coding genes in bloodstream-form trypanosomes, except those of the VSG gene expression sites, occurs by means of an RNA polymerase that is sensitive to α-amanitin at a concentration of 5 μg ml^{-1}. This concentration of α-amanitin, though slightly larger than that required for inhibition of transcription in higher eukaryotes, is characteristic of transcription by RNA polymerase II, which also transcribes the protein-coding genes of all other eukaryotes analyzed. Surprisingly, mRNA promoters with CAAT and TATA boxes at the appropriately spaced locations have so far not been found either in the intergenic regions or upstream of the tandem arrays of the RNA polymerase II-transcribed protein-coding genes. The only well-described RNA polymerase II transcription initiation site is that of the medRNA genes (194,195,197,199). Even in case of the medRNA genes, the characterization of the promoter and its comparison to other eukaryotic promoters has not led to an unambiguous identification of the control elements.

VSG genes and their expression sites are transcribed by an RNA polymerase that, in contrast to RNA polymerase II, is resistant to the drug α-amanitin at concentrations indicative of RNA polymerase I transcription (202). A single point mutation has been described which renders mouse RNA polymerase II resistant to this drug (231–233). However, the nucleotide sequence of the RNA polymerase II genes in trypanosomes is unaltered at this position, therefore failing to support the possibility that a modified RNA polymerase II transcribes the expression sites. More importantly, only one α-amanitin-resistant RNA polymerase has so far been identified by chromatographic techniques (see Section 5.2.1). It is thus possible that VSG genes, unlike protein-coding genes in any other eukaryote, are transcribed by RNA polymerase I. Alternatively, a modified RNA polymerase II or a novel transcription factor conferring α-amanitin resistance could explain the VSG gene transcription.

Because of the problems in identifying the transcription initiation sites of VSG genes, nuclear run-on assays have been performed to circumvent these difficulties. Shea and Van der Ploeg (102) analyzed the active expression sites of the 1.8 VSG gene located on a 430 kb chromosome and the 118 VSG gene located on a 1.5 Mb chromosome. They identified a small polygenic transcription unit of the VSG 118 gene which measures 7 kb (158). The 7000 nt polygenic pre-mRNA presumably matured via *trans*-splicing with the medRNA, generating several mature mRNAs in addition to the VSG mRNA. Immediately upstream of the 5' end of the polygenic pre-mRNA, a potential VSG gene transcription initiation site was identified whose nucleotide sequence resembled the rRNA promoter of *T. brucei* (158,234). Since at least one ESAG located upstream of the putative VSG gene promoter was also transcribed by the α-amanitin-insensitive RNA polymerase, at least two transcription units may be present in this VSG gene expression site. However, mapping transcription initiation sites by UV inactivation has so far failed to identify this RNA polymerase I-like promoter as an independent promoter in the expression site (M.Crozatier, P.Borst, L.H.T.Van der Ploeg et al., unpublished data). In a similar analysis, Kooter et al. (101) characterized the expression site with the 221 VSG gene located on a 3 Mb chromosome. The entire 55 kb locus of this expression site was transcribed by an α-amanitin-insensitive RNA polymerase. They located a single potential transcription initiation site at 55 kb upstream of the VSG gene, which has not yet been analyzed further (101). Pays *et al.* subsequently located a potential transcription initiation site in the expression site of variant AnTat 1.3, which resides 13 kb upstream of the VSG gene (103). This putative initiation site also shows homology to the trypanosomal rRNA promoter. Together, these results show that the VSG gene expression sites are similarly organized into polygenic transcription units and that the sites are transcribed by an RNA polymerase which differs from the classical RNA polymerase II. This RNA polymerase may be an RNA polymerase I. Rigorous testing of the identity of the RNA polymerase requires the isolation of the α-amanitin-resistant RNA polymerase and detailed analysis of its transcription initiation sites.

Transcription of expression sites by RNA polymerase I may appear implausible since most eukaryotic mRNAs require a 5' cap and pol I-derived RNAs are uncapped. However, all 5' ends of mature mRNAs in trypanosomes are indistinguishable because of the post-transcriptional addition, in *trans*, of the 5'-capped mini-exon.

6. Trypanosomes in the insect vector and expression site transcriptional control

When bloodstream-form trypanosomes are ingested by an insect vector, they differentiate into procyclic (insect-form) trypanosomes which no longer transcribe their VSG gene expression sites (162, 167–170). Hence these parasites appear 'coat-less'. In all cases analyzed, the loss of transcription at the expression site is a rapid event taking less than 24 h (235,236). The VSG mRNA half-life was also measured under these conditions and was shown to decrease dramatically by 4-fold, leading to the rapid loss of VSG mRNA. It is even possible *in vitro* to reduce and eliminate transcription at the expression sites by changing the temperature at which the bloodstream-form parasites are maintained from 37 to 27°C (103,202). This temperature shift mimics the temperature shift experienced *in vivo* since the insect-form of the parasite grows in the non-temperature regulated (poikilothermic) insect host (237). It is therefore possible that temperature plays a central role in the process of VSG inactivation.

The VSG gene expression sites are not the only regions containing protein-coding genes which are transcribed by the unusual α-amanitin-resistant RNA polymerase (238,263). The α-amanitin-resistant RNA polymerase in insect-form trypanosomes produces the mRNAs encoding the procyclic acidic repetitive (cell surface) proteins (PARP) of the insect-form parasites (239–242). The expression of these genes is again transcriptionally controlled. The PARP mRNA is as abundant as the VSG mRNA (2–5% of poly[A]$^+$). A specialized α-amanitin-resistant RNA polymerase is therefore utilized to generate the abundant mRNAs encoding the cell surface proteins in both bloodstream- and insect-form trypanosomes. This unique specialization, employing a distinct RNA polymerase, may be required because of the unusually high abundance of the mRNAs involved.

Surprisingly, transcriptional control of RNA polymerase II-transcribed genes has not yet been observed in trypanosomes. All RNA polymerase II-transcribed genes that have been analyzed in trypanosomes thus far are constitutively expressed (201,219). The steady-state mRNA levels of these genes appear to be post-transcriptionally controlled. The α-amanitin-resistant RNA polymerases that transcribe (PARP) and VSG genes may therefore control a select class of genes, the expression of which requires transcriptional control to allow for the rapid production of the abundant mRNAs.

6.1 Procyclic trypanosomes and DNA recombinational events

Procyclic trypanosomes undergo a variety of metabolic changes, among which is an absolute block in transcription of their VSG gene expression sites (243,244). The mechanisms of inactivation and reactivation of the expression sites have been studied. In a few examples, the expression site with the VSG gene was compared before and after the differentiation event. In most cases, the VSG gene and the expression site were retained and inactivated *in situ* (162,167 – 170). However, on a few rare occasions, the VSG gene of the expression site was lost (127). Pays and co-workers identified a different DNA recombinational event in which a short stretch of DNA, located just upstream of the VSG gene in the expression site, was deleted (139). This DNA sequence, named the companion, had previously been placed in the expression site by duplicative transposition and flanked the 5′ end of the VSG gene. A related companion sequence flanked the rDNA-like promoter of VSG gene 118 in variant 118 clone 1 (133,158). This companion sequence was also transposed into the expression site of trypanosome variant 118 clone 1 in an independent duplicative transposition event which presumably preceded the duplicative transposition of VSG gene 118. The function of the companion sequence and the occurrence of its deletion in expression site inactivation remains unclear.

Agabian and co-workers found that a different set of DNA sequences are deleted in the transition from bloodstream-form to insect-form trypanosomes (169). They studied the medRNA genes of trypanosomes. The medRNA genes exist as approximately 200 virtually identical genes arranged in tandem arrays. In *Trypanosoma gambiense*, the medRNA genes are located at two separate loci. Seven of the medRNA genes in *T.gambiense* are interrupted by virtually identical retroposons (245). The retroposon presumably integrated only once into a primordial medRNA gene and was subsequently carried along in the medRNA gene family through unequal sister chromatid exchange. Restriction enzyme digestion of the tandem array of medRNA genes generates uniformly sized restriction enzyme fragments derived from the tandem arrays of medRNA genes, as well as different sized polymorphic restriction enzyme fragments present as a result of the sites at which the retroposon integrated. Several of these polymorphic restriction enzyme fragments of *T.brucei* are lost during differentiation of the bloodstream-form into the insect-form. The mechanisms of these DNA rearrangement events have not been studied, but the fact that the fragments contain a retroposon is intriguing. A plausible explanation for this observation may involve the excision of retroposon sequences.

6.2 Metacyclic trypanosomes

The procyclic trypanosomes which are found in the hindgut of the insect vector differentiate into metacyclic trypanosomes in the salivary glands of the insect (246 – 252). These trypanosomes express a limited repertoire of approximately 14 different metacyclic VSG (mVAT) genes. This repertoire of mVATs is conserved when compared in different trypanosome isolates, though minor variations in the number of genes in the repertoire can be detected. The antigen switch rate in the metacyclic trypanosomes has been shown to be as high as 1 for every 30 cell divisions (250). The control of the mVAT expression appears to be at the transcriptional level (251).

The preferential activation of a fixed repertoire of VSGs in the metacyclic trypanosomes which populate the salivary glands indicates that these mVAT genes, like the PARP genes, are activated in a programmed stage-specific manner. Upon injection of these trypanosomes into the bloodstream of a mammal, the parasite rapidly switches from the expression of a fixed repertoire of mVATs to the expression of a repertoire of bloodstream VSGs. The first bloodstream variants that arise result from the preferential activation of the expression site that was inactivated upon ingestion by the insect vector.

Since the metacyclic trypanosomes cannot be maintained stably in culture, the mechanisms that are involved in switching and determining their preferential activation have not been studied in detail. However, a few interesting and relevant findings have emerged. First, some of the VSGs that are expressed in the metacyclic repertoire can reappear at later stages of the infection as part of the bloodstream VSG repertoire. Second, the VSG genes of mVATs are always found located at telomeres. The mVATs might therefore reside permanently within their own expression sites. Finally, the mVAT telomeres are not preceded by the tandem arrays of 70 bp repeats that flank duplicatively transposed VSG genes. The absence of these repeated DNA sequences in metacyclic expression sites may lead to a decrease in the frequency of gene replacement and hence to the stabilization of the metacyclic VSG repertoire (249,251).

The mechanisms that control the preferential activation and the relatively high switch rates occurring early in the infection are not understood. However, a high initial switch rate may ensure the establishment of a sufficiently diverged population of trypanosomes thus allowing the 'super-infection' of a host.

7. Evolution of the VSG gene repertoire

The repertoire of VSG genes is not stable and as a result, different geographic isolates of the same trypanosome species have markedly different repertoires. A rapid evolution of the pool of VSG genes may facilitate the co-existence of divergent trypanosome populations in the same geographic area. On the basis of nucleotide sequence homologies between C termini of the VSGs it has been shown that the VSGs can be grouped into families of genes (24–26,147). At the DNA level these families are not directly physically linked and are topologically mixed throughout the clustered arrays of VSG genes (21). The mechanisms involved in both the evolution and the topological mixing of VSG genes located at internal positions within the chromosomes and in the relatively high rate of mutation of telomeric VSG genes will be discussed below.

7.1 Unequal sister chromatid exchange and retroposition

The rate of evolution of several BC VSG gene loci was indirectly measured by Borst and co-workers (253,254). They analyzed the VSG gene 118 locus by physical mapping and compared the VSG 117, 121 and 221 loci in four related trypanosome stocks. From this analysis, they concluded that a subset of VSG genes evolves rapidly and developed a model for this evolutionary pattern stressing local hypermutagenesis.

When comparing the physical maps of DNA surrounding the VSG 118 gene in different *T.brucei* isolates, they identified modifications. First, the deletion of an entire VSG pseudogene and the insertion at this site of a TAA repeat. Second, they provided evidence for the expansion of a tandem array of VSG genes also containing VSG gene 118 by unequal sister chromatid exchange (255).

The family members of diverged VSGs are topologically mixed throughout the VSG loci indicating that unequal sister chromatid exchange may account for the dispersal of VSG genes in the genome. Alternative explanations of gene transfer to account for the topologically mixed appearance of VSG families might involve retroposition or, alternatively, gene conversion. However, the retroposition of VSG genes would be expected to spread genes throughout the genome at random. Since the tandem arrays of BC VSG genes are present on only about four of the 20 bands with chromosomes (21,63), unequal sister chromatid exchange in combination with inter- and intrachromosomal gene conversions is a

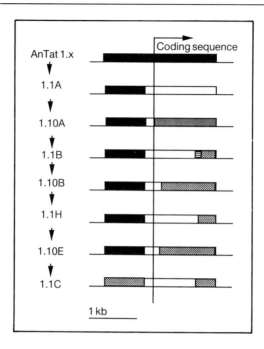

Figure 18.   Schematic representation of the successive changes in the sequence of the expressed antigen gene in different *T.brucei* clones (modified from ref. 134). The black box represents the companion sequence. The white, hatched, and shaded regions represent coding sequences derived from three different donor VSG genes.

more likely explanation for the chromosomal organization of the VSG genes. Evidence for the transfer of genetic information from telomeric sites to internal positions (e.g. by gene conversion) is lacking.

Compelling evidence for retroposition as a mechanism for gene transfer accumulated from the analysis of the trypanosome genome. These retroposition events do *not*, however, involve VSG gene transposition. They were characterized by retroposon hallmarks, which are the duplication of the target site and the presence of a poly[A] tail. The retroposons that have been characterized are:

(i)   the RIME retroposon, identified in a rDNA gene (256) and characterized by a 7 bp target site duplication and a 15 bp poly[A] tail;

(ii)   the INGE retroposon, sharing substantial homology with the human LINE retroposons (257). The Inge retroposon has complex termini, which are RIME derived; and

(iii)   the SLACS retroposon which interrupts several medRNA genes (245) and is characterized by a 49 bp target site duplication and a 36 bp terminal poly[A] tail.

7.2 Segmental gene conversion

The evolution of the repertoire of telomeric VSG genes occurs more rapidly than that of internal VSG genes (253,254). Several different lines of evidence indicate that segmental gene conversion is one of the processes involved in the modification of the telomeric VSG gene repertoire (123,134,138,258). The DNA sequence analysis of several related telomeric VSG genes indicates that these VSG genes differ by multiple point mutations (259). A 'sloppy' DNA polymerase was at first proposed as an explanation for the high rate of mutations. However, the mutation rate introduced by this polymerase would have to be at least 4% (253). The fact that almost all telomeric VSG genes are faithfully copied indicates that segmental gene conversion is a more likely and plausible explanation for the accumulated mutations. Pays and co-workers have shown that the segmental gene conversion of telomeric VSG genes can indeed account for the accumulated changes in the DNA sequence (134,138). In their study, three related VSG genes were compared. Exchanges ranging from several to hundreds of basepairs had occurred among the three genes (*Figure 18*).

The repertoire of telomeric VSG genes may thus undergo frequent gene conversion events, accounting for the rapid changes that are observed in the coding sequences of these genes.

8. Concluding remarks

The list of novel molecular mechanisms that have been discovered in trypanosomes is likely to increase and may very well continue both to enlighten and modify the existing dogmas of molecular genetics. In this way basic research on trypanosomes has added to our understanding of other systems in higher eukaryotes and may, in the future, lead to methods by which the detriment caused by these parasites can be prevented.

9. Acknowledgements

I thank Steven Brown, Keith Gottesdiener, Sylvia LeBlancq, Mary Gwo-Shu Lee, David Misek, Alan Morrison, Alison Rattray, Cathy Shea, and Michael Weiden for critical reading of the manuscript, and Drs Piet Borst and Albert Cornelissen for communicating unpublished results.

10. References

1. Borst, P. and Greaves, D. R. (1987) Programmed gene rearrangements. *Science,* **235**, 658.
2. Clayton, C. E. (1987) The molecular biology of Kinetoplastidae. *Gen. Engin.,* **7**, 1.

3. Boothroyd, J. C. (1985) Antigenic variation in African trypanosomes. *Annu. Rev. Microbiol.,* **39**, 475.
4. Donelson, J. E. and Rice-Ficht, A. C. (1985) Molecular biology of trypanosome antigenic variation. *Microbiol. Rev.,* **49**, 107.
5. Parsons, M., Nelson, R. G., and Agabian, N. (1984) Antigenic variation in African trypanosomes: DNA rearrangements program immune evasion. *Immunol. Today,* **5**, 43.
6. Van der Ploeg, L. H. T. (1987) Control of variant surface antigen switching in trypanosomes. *Cell,* **51**, 159.
7. Borst, P. (1986) Discontinuous transcription and antigenic variation in trypanosomes. *Annu. Rev. Biochem.,* **55**, 701.
8. Van der Ploeg, L. H. T. (1986) Discontinuous transcription and splicing in trypanosomes. *Cell,* **47**, 479.
9. Eisen, H. (1988) RNA editing: who's on first? *Cell,* **53**, 331.
10. Vickerman, K. (1978) Antigenic variation in trypanosomes. *Science,* **273**, 613.
11. Lumsden, W. H. R. and Evans, D. E. (eds) (1987) *Biology of the Kinetoplastida.* Academic Press, London.
12. Benne, R. (1986) Mitochondrial genes in trypanosomes. *Trends Genet.,* **1**, 117.
13. Simpson, L. (1986) Kinetoplast DNA in trypanosomid flagellates. *Int. Rev. Cytol.,* **99**, 119.
14. Sogin, M. L., Elwood, H. J., and Günderson, J. H. (1986) Evolutionary diversity of eukaryotic small-subunit rRNA genes. *Proc. Natl. Acad. Sci. USA,* **83**, 1383.
15. Vickerman, K. J. (1969) On the surface coat and flagellar adhesion in trypanosomes. *Cell Sci.,* **5**, 163.
16. Cross, G. A. M. (1975) Identification, purification and properties of clone specific glycoprotein antigens constituting the surface coat of *Trypanosoma brucei. Parasitology,* **71**, 393.
17. Cross, G. A. M. (1978) Antigenic variation in trypanosomes. *Proc. R. Soc. London Ser. B.,* **202**, 55.
18. Holder, A. A. and Cross, G. A. M. (1981) *Mol. Biochem. Parasitol.,* **2**, 135.
19. Ross, R. and Thomson, D. (1910) A case of sleeping sickness studied by precise enumerative methods: regular periodical increase of the parasites disclosed. *Proc. R. Soc. Lond. B. Biol. Sci.,* **82**, 411.
20. Capbern, A., Giroud, C., Baltz, T., and Mattern, P. (1977) *Trypanosoma equiperdum:* etude des variations antigeniques au cours de la trypanosomose experimentale du lapin. *Exp. Parasitol.,* **42**, 6.
21. Van der Ploeg, L. H. T., Valerio, D., De Lange, T., Bernards, A., Borst, P., and Grosveld, F. G. (1982) An analysis of cosmid clones of nuclear DNA from *Trypanosoma brucei* shows that the genes for variant surface glycoproteins are clustered in the genome. *Nucleic Acids Res.,* **10**, 5905.
22. Hoare, C. A. and Wallace, F. G. (1966) Developmental stages of trypanosomatid flagellates: a new terminology. *Nature,* **212**, 1385.
23. Boothroyd, J. C., Paynter, C. A., Cross, G. A. M., Bernards, A., and Borst, P. (1981) Variant surface glycoproteins of *Trypanosoma brucei* are synthesized with cleavable hydrophobic sequences at the carboxy and amino terminus. *Nucleic Acids Res.,* **9**, 4735.
24. Matthijssens, G., Michiels, F., Hamers, R., Pays, E., and Steinert, M. (1981) Two variant surface glyocproteins of *Trypanosoma brucei* have a conserved C-terminus. *Nature,* **293**, 230.
25. Rice-Ficht, A. C., Chen, K. K., and Donelson, J. E. (1981) Sequence homologies near the C-termini of the variable surface glycoproteins of *Trypanosoma brucei. Nature,* **294**, 53.
26. Boothroyd, J. C., Cross, G. A. M., Hoeijmakers, J. H. J., and Borst, P. (1980) A variant surface glycoprotein of *Trypanosoma brucei* synthesized with a C-terminal hydrophobic 'tail' absent from purified glycoprotein. *Nature,* **288**, 624.
27. Cross, G. A. M. (1987) Eukaryotic protein modification and membrane attachment via phosphatidylinositol. *Cell,* **48**, 179.
28. Cardoso de Almeida, M. L., Allan, L. M., and Turner, M. J. (1984) Purification and properties of the membrane form of VSGs from *Trypanosoma brucei. J. Protozool.,* **31**, 53.
29. Ferguson, M. A. J. and Cross, G. A. M. (1984) Myristylation of the membrane form of a *Trypanosoma brucei* variant surface glycoprotein. *J. Biol. Chem.,* **259**, 3011.
30. Cardosa de Almeida, M. L. and Turner, M. J. (1983) The membrane form of variant

surface glycoproteins of *Trypanosoma brucei*. *Nature,* **302**, 349.

31. Ferguson, M. A. J., Haldar, K., and Cross, G. A. M. (1985) *Trypanosoma brucei* variant surface glycoprotein has a sn-1,2-dimyristyl glycerol membrane anchor at its COOH terminus. *J. Biol. Chem.,* **260**, 4963.

32. Ferguson, M. A. J., Low, M. G., and Cross, G. A. M. (1985) Glycosyl-sn-1,2-dimyristyl-phosphatidylinositol is covalently linked to *Trypanosoma brucei* variant surface glycoprotein. *J. Biol. Chem.,* **260**, 14547.

33. Ferguson, M. A. J., Homans, S. W., Dwek, R. A., and Rademacher, T. W. (1988) Glycosyl-phosphatidylinositol moiety that anchors *Trypanosoma brucei* variant surface glycoprotein to the membrane. *Science,* **239**, 753.

34. Masterson, W. J., Doering, T. L., Hart, G. W., and Englund, P.T. A novel pathway for glycan assembly: biosynthesis of the glycosyl-phosphatidylinositol anchor of the trypanosome variant surface glycoprotein. *Cell,* **56**, 793.

35. Fox, J. A., Duszenkof, M., Ferguson, M. A. J., Low, M. G., and Cross, G. A. M. (1986) Purification and characterization of a novel glycan phosphatidylinositol-specific phospholipidase C from *Trypanosoma brucei*. *J. Biol. Chem.,* **261**, 15767.

36. Grab, D. J., Webster, P., Ito, S., Fish, W. R., Verjee, Y., and Lonsdale-Eccles, J. D. (1987) Subcellular localization of a variable surface glycoprotein phosphatidylinositol-specific phospholipase-C in African trypanosomes. *J. Cell Biol.,* **105**, 737.

37. Metcalf, P., Blum, D., Freymann, D., Turner, M., and Wiley, D. C. (1987) Two variant surface glycoproteins of *Trypanosoma brucei* of different classes have similar 6 Å resolution X-ray structures. *Nature,* **325**, 84.

38. Donelson, J. E. and Turner, M. J. (1985) How the trypanosome changes its coat. *Sci. Am.,* **252**, 44.

39. McLaughlin, J. (1987) *Trypanosoma rhodosiense*: antigenicity and immunogenicity of flagellar pocket membrane components. *Exp. Parasitol.,* **64**, 1.

40. Coppens, I., Opperdoes, F. R., Courtoy, P. J., and Baudhuin, P. (1987) Receptor-mediated endocytosis in the bloodstream form of *Trypanosoma brucei*. *J. Protozool.,* **34**, 465.

41. Coppens, I., Baudhuin, P., Opperdoes, F. R., and Courtoy, P. J. (1988) Receptors for the host low-density lipoproteins on the haemoflagellate *Trypanosoma brucei*: purification and involvement in the growth of the parasite. *Proc. Natl. Acad. Sci. USA,* **85**, 6753.

42. Borst, P., Van der Ploeg, M., Van Hoek, J. F. M., Tas, J., and James, J. (1982) On the DNA content and ploidy of trypanosomes. *Mol. Biochem. Parasitol.,* **6**, 13.

43. Borst, P., Fase-Fowler, F. F., Frasch, A. C. C., Hoeijmakers, J. H. J., and Weijers, P. J. (1980) Characterization of DNA from *Trypanosoma brucei* and related trypanosomes by restriction endonuclease digestion. *Mol. Biochem. Parasitol.,* **1**, 221.

44. Gibson, W., Osinga, K. A., Michels, P. A. M., and Borst, P. (1985) Trypanosomes of the subgenus trypanozoon are diploid for housekeeping genes. *Mol. Biochem. Parasitol.,* **16**, 231.

45. Richardson, J. P., Jenni, L., Beecroft, R. P., and Pearson, T. W. (1986) Procyclic tsetse fly midgut forms and culture forms of African trypanosomes share stage- and species-specific surface antigens identified by monoclonal antibodies. *J. Immunol.,* **136**, 2259.

46. Steiger, R. F. (1973) On the ultrastructure of *Trypanosoma (Trypanozoon) brucei* in the course of its life cycle and some related aspects. *Acta Trop. (Basel),* **30**, 64.

47. Honigberg, B. M., Cunningham, I., Stanlye, H. A., Su-Lin, K. E., and Luckins, A. G. (1976) *Trypanosoma brucei* antigenic analysis of vector, and culture stages by the quantitative fluorescent antibody methods. *Exp. Parasitol.,* **39**, 496.

48. Crowe, J. S., Barry, J. D., Luckins, A. G., Ross, C. A., and Vickerman, K. (1983) All metacyclic variable antigen types of *Trypanosoma congolense* identified using monoclonal antibodies. *Nature,* **306**, 389.

49. Tait, A. (1983) Sexual processes in Kinetoplastida. *Parasitology,* **86**, 29.

50. Tait, A., Babiker, E. A., and Le Ray, D. (1984) Enzyme variation in *Trypanosoma brucei* spp. Evidence for the sub-speciation of *Trypanosoma brucei gambiense*. *Parasitology,* **89**, 311.

51. Zampetti-Bosseler, F., Schweizer, J., Pays, E., Jenni, L., and Steinert, M. (1986) Evidence for haploidy in metacyclic forms of *Trypanosoma brucei*. *Proc. Natl. Acad. Sci. USA,* **83**, 6063.

52. Paindavoine, P., Zampetti-Bosseler, F., Pays, E., Schweizer, J., Guyaux, M., Jenni, L., and Steinert, M. (1986) Trypanosome hybrids generated in tsetse flies by nuclear fusion. *EMBO J., 5*, 3631.

53. Jenni, L., Marti, S., Schweizer, J., Betschart, B., Le Page, R. W. F., Wells, J. M., Tait, A., Paindavoine, P., Pays, E., and Steinert, M. (1986) Hybrid formation between African trypanosomes during cyclical transmission. *Nature, 322*, 173.

54. Schweizer, J., Tait, A., and Jenni, L. (1988) The timing and frequency of hybrid formation in African trypanosomes during cyclical transmission. *Parasitol. Res., 75*, 98.

55. Le Page, R. W. F., Wells, J. M., Prospero, T. D., and Sternberg, J. (1988) Genetic analysis of hybrid *T. brucei*. In *Current Communications in Molecular Biology: Molecular Genetics of Parasitic Protozoa*. Turner, M. J. and Arnot, D. (eds), Cold Spring Harbor Laboratory, Cold Spring Harbor, NY, p. 65.

56. Tait, A., Sternberg, J., and Turner, C. M. R. (1988) Genetic exchange in *Trypanosoma brucei*: allelic segregation and re-assortment. In *Current Communications in Molecular Biology: Molecular Genetics of Parasitic Protozoa*. Cold Spring Harbor Laboratory, Cold Spring Harbor, NY p. 58.

57. Schwartz, D. and Cantor, C. R. C. (1984) Separation of yeast chromosome-sized DNAs by pulsed field gradient gel electrophoresis. *Cell, 37*, 67.

58. Van der Ploeg, L. H. T., Schwartz, D., Cantor, C. R., and Borst, P. (1984) Antigenic variation in *Trypanosoma brucei* analyzed by electrophoretic separation of chromosome-sized DNA molecules. *Cell, 37*, 77.

59. Carle, G. F. and Olson, M. V. (1985) An electrophoretic karyotype for yeast. *Proc. Natl. Acad. Sci. USA, 82*, 3756.

60. Carle, G. F. and Olson, M. V. (1984) Separation of chromosomal DNA molecules from yeast by orthogonal-field-alternation gel electrophoresis. *Nucleic Acids Res., 12*, 5647.

61. Bernards, A., Kooter, J. M,. Michels, P. A. M., Moberts, R. M. P., and Borst, P. (1986) Pulsed field gradient gel-electrophoresis of DNA digested in agarose allowing the sizing of the large duplication unit of a surface antigen gene in trypanosomes. *Gene, 42*, 313.

62. Smith, C. L. and Cantor, C. R. (1987) Purification, specific fragmentation and separation of large DNA molecules. *Methods Enzymol., 155*, 449.

63. Van der Ploeg, L. H. T., Smith, C. L., Polvere, R. I., and Gottesdiener, K. (1989) Improved separation of chromosome-sized DNA from *Trypanosoma brucei*, stock 427-60. *Nucleic Acids Res., 17*, 3217.

64. Borst, P., Bernards, A., Van der Ploeg, L. H. T., Michels, P. A. M., Liu, A. Y. C., De Lange, T., Sloof, P., Schwartz, D., and Cantor, C. R. C. (1983) The role of mini-chromosomes and gene translocation in the expression and evolution of VSG genes. In *Gene Expression*, UCLA Symposium on Molecular and Cellular Biology, Vol. 8. Hamer,D. and Rosenberg, H. (eds), Alan R.Liss, New York, p. 413.

65. Van der Ploeg, L. H. T. (1987) Separation of chromosome-sized DNA molecules by pulsed field gel electrophoresis. *Am. Biotechnol. Lab., 5*, 8.

66. Williams, R. O., Young, J. R., and Majiwa, P. A. O. (1982) Genomic environment of *T. brucei* VSG genes: presence of a mini-chromosome. *Nature, 299*, 417.

67. Rothwell, V., Aline, R., Jr, Parasons, M., Agabian, N., and Stuart, K. (1985) Expression of a minichromosomal variant surface glycoprotein in *Trypanosoma brucei*. *Nature, 313*, 595.

68. Majiwa, P. A. O., Young, J. R., Hamers, R., and Matthyssens, G. (1986) Mini-chromosomal variable surface glycoprotein genes and molecular karyotypes of *Trypanosoma (Nannomonas) congolense*. *Gene, 41*, 183.

69. Van der Ploeg, L. H. T., Cornelissen, A. W. C. A., Barry, J. D., and Borst, P. (1984) Chromosomes of Kinetoplastida. *EMBO J., 3*, 3109.

70. Van der Ploeg, L. H. T., Cornelisson, A. W. C. A., Michels, P. A. M., and Borst, P. (1984) Chromosome rearrangements in *Trypanosoma brucei*. *Cell, 39*, 213.

71. Van der Ploeg, L. H. T. and Cornelissen, A. W. C. A. (1984) The contribution of chromosomal translocation to antigenic variation in *Trypanosoma brucei*. *Phil. Trans. R. Soc. Lond. B, 307*, 13.

72. Van der Ploeg, L. H. T., Smits, M., Ponnudurai, T., Vermuelen, A., Meuwissen, J. H. E. Th., and Langsley, G. (1985) Chromosome-sized DNA molecules of *Plasmodium falciparum*. *Science, 16*, 658.

73. Kemp, D. J., Corcoran, L. M., Coppel, R. L., Stahl, H. D., Bianco, A. E., Brown,

G. V., and Anders, R. F. (1985) Size variation in chromosomes from independent cultured isolates of *Plasmodium falciparum. Nature,* **315**, 347.

74. Giannini, S. H., Schittini, M., Keithly, J. S., Warburton, P., Cantor, C. R., and Van der Ploeg, L. H. T. (1986) Karyotype analysis of *Leishmania* species and its use in classification and clinical diagnosis. *Science,* **232**, 761.

75. Scholler, J. K., Reed, S. G., and Stuart, K. (1986) Molecular karyotype of species and subspecies of *Leishmania. Mol. Biochem. Parasitol.,* **20**, 279.

76. Spithill, T. W. and Samaraa, N. (1985) The molecular karyotype of *Leishmania* major and mapping of alpha- and beta-tubulin gene families to multiple unlinked chromosomal loci. *Nucleic Acids Res.,* **13**, 4155.

77. Bishop, R. P. and Miles, M. A. (1987) Chromosome size polymorphisms of *Leishmania donovani. Mol. Biochem. Parasitol.,* **24**, 263.

78. Pologe, L. G. and Ravetch, J. A. (1986) A chromosomal rearrangement in a *P.falciparum* histidine-rich protein gene is associated with the knobless phenotype. *Nature,* **322**, 474.

79. Corcoran, L. M., Thompson, J. K., Walliker, D., and Kemp, D. (1988) Homologous recombination within subtelomeric repeat sequences generates chromosome size polymorphisms in *P.falciparum Cell,* **53**, 807.

80. Van der Ploeg, L. H. T., Liu, A. Y. C., and Borst, P. (1984) The structure of the growing telomeres of trypanosomes. *Cell,* **36**, 459.

81. Blackburn, E. H. and Challoner, P. B. (1984) Identification of a telomeric DNA sequence in *Trypanosoma brucei. Cell,* **36**, 447.

82. Blackburn, E. H. (1984) Telomeres: do the ends justify the means? *Cell,* **37**, 7.

83. Blackburn, E. H. (1986) Molecular structure of telomeres in lower eukaryotes. In *Molecular Developmental Biology.* Bogorad,L. (ed.), Alan R. Liss, New York, Inc., p. 69.

84. Blackburn, E. H. and Szostak, J. W. (1983) The molecular stucture of centromeres and telomeres. *Annu. Rev. Biochem.,* **53**, 163.

85. Lima-de-Faria, A. (1983) Organization and function of telomeres. In *Molecular Evolution and Organization of the Chromosomes.* Elsevier Science Publishers, p. 1.

86. Lima-de-Faria, A. (1983) The organization of the nuclear chromosome. The chromosome field. In *Molecular Evolution and Organization of the Chromosome.* Elsevier Science Publishers, Amsterdam, p. 329.

87. Morin, G. B. and Cech, T. R. (1988) Mitochondrial telomeres: surprising diversity of repeated telomeric DNA sequences among six species of *Tetrahymena. Cell,* **52**, 367.

88. Bernards, A., Michels, P. A. M., Lincke, C. R., and Borst, P. (1983) Growth of chromosome ends in multiplying trypanosomes. *Nature,* **303**, 592.

89. Greider, C. W. and Blackburn, E. (1987) The telomere terminal transferase of *Tetrahymena* is a ribonucleoprotein enzyme with two kinds of primer specificity. *Cell,* **51**, 887.

90. Henderson, E., Hardin, C. C., Walk, S. K., Tinoco, I., Jr, and Blackburn, E. (1987) Telomeric DNA oligonucleotides form novel intramolecular structures containing guanine–guanine base pairs. *Cell,* **51**, 899.

91. Larson, D. D., Spangler, E. A., and Blackburn, E. (1987) Dynamics of telomere length variation in *Tetrahymena thermophila. Cell,* **50**, 477.

92. Sloof, P., Menke, H. H., Caspers, M. P. M., and Borst, P. (1983) Size fractionation of *Trypanosoma brucei* DNA: localization of the 177 bp repeat satellite DNA and a variant surface glycoprotein gene in a mini-chromosomal DNA fraction. *Nucleic Acids Res.,* **11**, 3889.

93. Sloof, P., Bos, J. L., Konings, A. F. J. M., Menke, H. H., Borst, P., Gutteridge, W. E., and Leon, W. (1983) Characterization of satellite DNA in *Trypanosoma brucei* and *Trypanosoma cruzi. J. Mol. Biol.,* **167**, 1.

94. Solari, A. J. (1980) The 3-dimensional fine structure of the mitotic spindle in *Trypanosoma cruzi. Chromosoma (Berlin),* **78**, 239.

95. Solari, A. J. (1983) The ultrastructure of mitotic nuclei of *Blastocrithidia-triatomae. Z. Parasitenkund.,* **69**, 3.

96. Solari, A. J. (1982) Nuclear ultrastructure during mitosis in *Crithidia-fasciculata* and *Trypanosoma brucei brucei. J. Protozool.,* **29**, 330.

97. Doyle, J. J,. Hirumi, H., Hirumi, K., Lupton, E. N., and Cross, G. A. M. (1980) Antigenic variation in clones of animal infective *Trypanosoma brucei* derived and maintained *in vitro. Parasitology,* **80**, 359.

98. Lamont, G. S., Tucker, R. S., and Cross, G. A. M. (1986) Analysis of antigen switching

rates in *Trypanosoma brucei. Parasitology,* **92**, 355.

99. De Lange, T. and Borst, P. (1982) Genomic environment of the expression-linked extra copies of genes of surface antigens of *Trypanosoma brucei* resembles the end of a chromosome. *Nature,* **299**, 451.

100. Cully, D. F., Ip, H. S., and Cross, G. A. M. (1985) Coordinate transcription of variant surface glycoprotein genes and an expression site associated gene family in *Trypanosoma brucei. Cell,* **42**, 173.

101. Kooter, J. M., Van der Spek, H. J., Wagter, R., D'Oliveira, C. E., Van der Hoeven, F., Johnson, P., and Borst, P. (1987) The anatomy and transcription of a telomeric expression site for variant-specific surface antigens in *Trypanosoma brucei. Cell,* **51**, 261.

102. Shea, C. and Van der Ploeg, L. H. T. (1988) Stable variant specific transcripts of the variant cell surface glycoprotein gene 1.8 expression site in *Trypanosoma brucei. Mol. Cell. Biol.,* **8**, 854.

103. Alexandre, S., Guyaux, M., Murphy, N. B., Coquelet, H., Pays, A., Steinert, M., and Pays, E. (1988) Putative genes of a variant-specific antigen gene transcription unit in *Trypanosoma brucei. Mol. Cell. Biol.,* **8**, 2367.

104. Kooter, J. M., Winter, A. J., D'Oliveira, C., Wagter, R., and Borst, P. (1988) Boundaries of telomere conversion in *Trypanosoma brucei. Gene,* **69**, 1.

105. Gibbs, C. P. and Cross, G. A. M. (1988) Cloning and transcriptional analysis of a variant surface glycoprotein gene expression site in *Trypanosoma brucei. Mol. Biochem. Parasitol.,* **28**, 197.

106. Van der Ploeg, L. H. T., Shea, C., Polvere, R. I., and Lee, M. G. -S. (1987) Chromosomal rearrangements and VSG gene transcription in *Trypanosoma brucei.* In *Molecular Strategies of Parasite Invasion.* UCLA Symposia on Molecular and Cellular Biology, Vol. 42. Agabian,N., Goodman,H. and Noguiera,N. (eds), Alan R.Liss, Inc., New York, p. 437.

107. Van der Ploeg, L. H. T., Shea, C., Lee, M. G. -S., Huang, J., Morrison, A., Ralph, D., and Rudenko, G. (1988) Transcriptional control of VGS genes. In *Current Communications in Molecular Biology.* Turner,M.J. and Arnot,D. (eds), Cold Spring Harbor Laboratory Press, Cold Spring Harbor, NY, p. 77.

108. Cully, D. F., Gibbs, C. P., and Cross, G. A. M. (1987) Identification of proteins encoded by variant surface glycoprotein expression site associated genes in *Trypanosoma brucei. Mol. Biochem. Parasitol.,* **21**, 189.

109. Kooter, J. M., Winter, A. J., de Oliveira, C., Wagter, R., and Borst, P. (1988) Boundaries of telomere conversion in *Trypanosoma brucei. Gene,* **69**, 1.

110. Crowe, J. S., Barry, J. D., Luckins, A. G., Ross, C. A., and Vickerman, K. (1983) All metacyclic variable antigen types of *Trypanosoma congolense* identified using monoclonal antibodies. *Nature,* **306**, 389.

111. Cunningham, I. (1986) Infectivity of *Trypanosoma rhodesiense* cultivated at 28°C with various tsetse fly tissues. *J. Protozool.,* **33**, 226.

112. Cunningham, I. and Kaminsky, R. (1986) Development of metacyclic forms of *Trypanosoma brucei* spp. in cultures containing explants of *Phormia regina. J. Parasitol.,* **72**, 944.

113. Liu, A. Y. C., Michels, P. A. M., Bernards, A., and Borst, P. (1985) Trypanosome variant surface glycoprotein genes expressed early in infection. *J. Mol. Biol.,* **175**, 383.

114. Aline, R. F. J., Scholler, J. K., Nelson, R. G., Agabain, N., and Stuart, K. (1985) Preferential activation of telomeric variant surface glycoprotein in *Trypanosoma brucei. Mol. Biochem. Parasitol.,* **17**, 311.

115. Seed, J. R., Edwards, R., and Secheelski, J. (1984) The ecology of antigenic variation. *J. Protozool.,* **31**, 48.

116. Miller, E. N. and Turner, M. J. (1981) Analysis of antigenic types appearing in first relapse populations of clones of *Trypanosoma brucei. Parasitology,* **82**, 63.

117. De Gee, A. L. W., Shah, S., and Doyle, J. J. (1981) *Trypanosoma vivax*: host influence on appearance of variable antigen types. *Exp. Parasitol.,* **51**, 392.

118. De Gee, A. L. W., Shah, S. D., and Doyle, J. J. (1979) *Trypanosoma vivax*: sequence of antigenic variants in mice and goats. *Exp. Parasitol.,* **48**, 352.

119. Hoeijmakers, J. H. J., Frasch, A. C. C., Bernards, A., Borst, P., and Cross, G. A. M. (1980) Novel expression linked copies of the genes for variant surface antigens in trypanosomes. *Nature,* **284**, 78.

120. Bernards, A., Van der Ploeg, L. H. T., Frasch, A. C. C., Borst, P., Boothroyd, J.

C., Coleman, S., and Cross, G. A. M. (1981) Activation of trypanosome surface glycoprotein genes involves duplication – transcription leading to an altered 3' end. *Cell,* **27**, 497.

121. Van der Ploeg, L. H. T., Bernards, A., Rijsewijk, F. A. M., and Borst, P. (1982) Characterization of the DNA duplication – transposition that controls the expression of two genes for variant surface glycoproteins in *Trypanosoma brucei. Nucleic Acids Res.,* **10**, 593.

122. Pays, E., Lheureux, M., and Steinert, M. (1981) The expression linked copy of the surface antigen gene in *Trypanosoma brucei* is probably the one transcribed. *Nature,* **292**, 265.

123. Pays, E., Van Assel, A., Laurent, M., Darville, M., Vervoort, T., Van Meirvenne, N., and Steinert, M. (1983) Gene conversion as a mechanism for antigenic variation in trypanosomes. *Cell,* **34**, 371.

124. Young, J. R., Donelson, J. G., Majiwa, P. A. O., Shapiro, S. Z., and Williams, R. O. (1982) Analysis of genomic rearrangements associated with two variable antigen genes of *Trypanosoma brucei. Nucleic Acids Res.,* **10**, 803.

125. Pays, E., Lheureux, M., and Steinert, M. (1981) Analysis of the DNA and RNA changes associated with the expression of the isotypic variant-specific antigens of trypanosomes. *Nucleic Acids Res.,* **9**, 4225.

126. Pays, E., Lheureux, M., and Steinert, M. (1982) Structure and expression of a *Trypanosoma brucei gambiense* variant specific antigen gene. *Nucleic Acids Res.,* **10**, 3149.

127. Pays, E., Van Meirvenne, N., Le Ray, D., and Steinert, M. (1981) Gene duplication and transposition linked to antigenic variation in *Trypanosoma brucei Proc. Natl. Acad. Sci. USA,* **78**, 2673.

128. De Lange, T., Kooter, J. M., Michels, P. A. M., and Borst, P. (1983) Telomere conversion in trypanosomes. *Nucleic Acids Res.,* **11**, 8149.

129. Pays, E., Guyaux, M., Aerts, D., Van Meirvenne, N., and Steinert, M. (1986) Telomeric reciprocal recombination as a possible mechanism for antigenic variation in trypanosomes. *Nature,* **316**, 562.

130. Michels, P. A. M., Liu, A. Y. C., Bernards, A., Sloof, P., Van der Bijl, M. M. W., Schinkel, A. H., Menke, H. H., and Borst, P. (1983) Activation of the genes for variant surface glycoproteins 117 and 118 in *Trypanosoma brucei. J. Mol. Biol.,* **166**, 537.

131. Michels, P. A. M., Bernards, A., Van der Ploeg, L. H. T. and Borst, P. (1982) Characterization of the expression linked gene copies of variant surface glycoprotein 118 in two independently isolated clones of *Trypanosoma brucei. Nucleic Acids Res.,* **10**, 2353.

132. Timmers, H. Th. M. de Lange, T., Kooter, J. M., and Borst, P. (1987) Coincident multiple activations of the same surface antigen gene in *Trypanosoma brucei. J. Mol. Biol.,* **194**, 81.

133. Lee, M. G. -S. and Van der Ploeg, L. H. T. (1987) Frequent independent duplicative transpositions activate a single VSG gene. *Mol. Cell. Biol.,* **7**, 357.

134. Pays, E., Houard, S., Pays, A., Van Assel, S., Dupont, F., Aerts, D., Huet-Duvillier, G., Gomeo, V., Richet, C., Degand, P., Van Meirvenne, N., and Steinert, M. (1985) *Trypanosoma brucei*: the extent of conversion in antigen genes may be related to the DNA coding specificity. *Cell,* **42**, 821.

135. Pays, E. (1988) DNA recombination and transposition in trypanosomes. In *Transposition.* Society for General Microbiology Symposium 43, Kingsman, A. J., Kingsman, S. M., and Chater, K. F. (eds). Cambridge University Press, Cambridge, p. 301.

136. Pays, E. (1985) Gene conversion in trypanosome antigenic variation. *Prog. Nucleic Acid Res. Mol. Biol.,* **32**, 1.

137. Pays, E. (1986) Variability of antigen genes in African trypanosomes. *Trends Genet.,* **2**, 21.

138. Pays, E., Delauw, M. F., Van Assel, S., Laurent, M., Vervoort, T., Van Meirvenne, N., and Steinert, M. (1983) Modifications of a *Trypanosoma b. brucei* antigen gene repertoire by different DNA recombinational mechanisms. *Cell,* **35**, 721.

139. Pays, E., Van Assel, S., Laurent, M., Dero, B., Michiels, F., Kronenberger, P., Mathyssens, G., Van Meirvenne, N., Le Ray, D., and Steinert, M. (1983) At least two transposed sequences are associated in the expression site of a surface antigen gene in different trypanosome clones. *Cell,* **34**, 359.

140. Liu, A. Y. C., Van der Ploeg, L. H. T., Rijsewijk, F. A. M., and Borst, P. (1983)

The transcription unit of variant surface glycoprotein gene 118 of *Trypanosoma brucei.* Presence of repeated elements at its border and absence of promoter associated sequences. *J. Mol. Biol.,* **167**, 57.

141. Shah, J. S., Young, J. R., Kimmel, B. E., Iams, K. P., and Williams, R. O. (1987) The 5' flanking sequence of *Trypanosoma brucei* variable surface glycoprotein gene. *Mol. Biochem. Parasitol.,* **24**, 163.

142. Florent, I., Baltz, T., Raibaud, A., and Eisen, H. (1987) On the role of repeated sequences 5' to variant surface glycoprotein genes in African trypanosomes. *Gene,* **53**, 55.

143. Campbell, D. A., Van Bree, M. P., and Boothroyd, J. C. (1984 The 5'-limit of transposition and upstream-barren region of a trypanosome VSG gene: tandem 76 base-pair repeats flanking (TAA)90. *Nucleic Acids Res.,* **12**, 2759.

144. Bernards, A., de Lange, T., Michels, P. A. M., Liu, A. Y. C., Huisman, M. J., and Borst, P. (1984) Two modes of activation of a single surface antigen gene of *Trypanosoma brucei. Cell,* **36**, 163.

145. Myler, P. J., Nelson, R. G., Agabian, N., and Stuart, K. (1984) Two mechanisms of expression of a predominant variant antigen gene of *Trypanosoma brucei. Nature,* **309**, 282.

146. De Lange, T., Kooter, J. M., Luirink, J., and Borst, P. (1985) Transcription of a transposed trypanosome surface antigen gene starts upstream of the transposed segment. *EMBO J.,* **4**, 3299.

147. Borst, P., Frasch, A. C. C., Bernards, A., Van der Ploeg, L. H. T., Hoeijmakers, J. H. J., Arnberg, A., and Cross, G. A. M. (1981) DNA rearrangements involving the genes for variant antigens in *Trypanosoma brucei. Cold Spring Harbor Sympt. Quant. Biol.,* **45**, 935.

148. Borst, P., Bernards, A., Van der Ploeg, L. H. T., Michels, P. A. M., Liu, A. Y. C., De Lange, T. and Kooter, J. M. (1983) The control of variant surface antigen synthesis in trypanosomes. *Eur. J. Biochem.,* **137**, 383.

149. Borst, P., Bernards, A., Van der Ploeg, L. H. T., Michels, P. A. M., Liu, A. Y. C., and De Lange, T. (1983) Gene rearrangements controlling the expression of genes for variant surface antigens in trypanosomes. In *Tumor Viruses and Differentiation.* Alan R.Liss, New York, p. 243.

150. Shapiro, J. A. (ed) (1983) Mobile genetic elements. Academic Press, Inc., New York.

151. Weiner, A. M., Dininger, P. L., and Efstratiadis, A. (1986) Nonviral retroposons: genes, pseudogenes, and transposable elements generated by the reverse flow of genetic information. *Annu. Rev. Biochem.,* **55**, 631 – 661.

152. Kostriken, R., Strathern, J. N., Klar, A. J. S., Hicks, J. B., and Heffron, F. (1983) A site-specific endonuclease essential for mating-type switching in *Saccharomyces cerevisiae. Cell,* **35**, 167.

153. Nasmyth, K. A. (1982) The regulation of yeast mating type chromatin structure by SIR: an action at a distance affecting both transcription and transposition. *Cell,* **30**, 567.

154. Klar, A. J. S., Strathern, J. N., and Hicks, J. B. (1981) A position-effect control for gene transposition: state of expression of yeast mating type genes affects their ability to switch. *Cell,* **25**, 517.

155. Jackson, J. A. and Fink, G. R. (1981) Gene conversion between duplicated genetic elements in yeast. *Nature,* **292**, 306.

156. Reynaud, C. A., Anquez, V., Brimal, H., and Weill, J. C. (1987) A hyperconversion mechanisms generates the chicken light chain preimmune repertoire. *Cell,* **48**, 379.

157. Johnson, P., Kooter, J. M., and Borst, P. (1987) Inactivation of transcription by UV irradiation of *Trypanosoma brucei* provides further evidence for a multicistronic transcription unit that includes a variant surface glycoprotein gene. *Cell,* **51**, 273.

158. Shea, C., Lee, M. G. -S. and Van der Ploeg, L. H. T. (1987) VSG gene 118 is transcribed from a co-transposed pol I-like promoter. *Cell,* **50**, 603.

159. Young, J. R. and Majiwa, P. A. O. (1979) Genomic rearrangements correlated with antigenic variation in *Trypanosoma brucei. Nature,* **282**, 847.

160. Young, J. R., Shah, J. S., Matthyssens, G., and Williams, R. O. (1983) Relationship between multiple copies of a *T. brucei* variable surface glycoprotein gene whose expression is not controlled by duplication. *Cell,* **32**, 1149.

161. Laurent, M., Pays, E., Magnus, E., Van Meirvenne, N., Mathyssens, G., Williams, R. O., and Steinert, M. (1983) DNA rearrangements linked to the expression of a

predominant surface antigen gene of trypanosomes. *Nature*, **302**, 263.

162. Buck, G. A., Jacquemot, C., Baltz, T., and Eisen, H. (1984) Reexpression of an inactivated varible surface glycoprotein gene in *Trypanosoma equiperdum*. *Gene*, **32**, 329.

163. Longacre, S., Hibner, U., Raibaud, A., Eisen, H., Baltz, T., Giroud, C., and Baltz, D. (1983) DNA rearrangements and antigenic variation in *Trypanosoma equiperdum*: multiple expression linked sites in independent isolates of trypanosomes expressing the same antigen. *Mol. Cell. Biol.*, **3**, 399.

164. Myler, P. J., Allison, J., Agabian, N., and Stuart, K. (1984) Antigenic variation in African trypanosomes by gene preplacement or activation of alternate telomeres. *Cell*, **39**, 203.

165. Bernards, A. (1982) Transposable genes for surface glycoproteins in trypanosomes. *Trends Biochem. Sci.*, **7**, 253.

166. Bernards, A., Kooter, J. M., and Borst, P. (1985) Structure and transcription of a telomeric surface antigen gene of *Trypanosoma brucei*. *Mol. Cell. Biol.*, **5**, 545.

167. Michels, P. A. M., Van der Ploeg, L. H. T., Liu, A. Y. C. and Borst, P. (1984) The inactivation and reactivation of an expression-linked gene copy for a variant surface glycoprotein in *Trypanosoma brucei*. *EMBO J.*, **3**, 1345.

168. Michiels, F., Matthyssens, G., Kronenberger, P., Pays, E., Dero, B., Van Assel, A., Darville, M., Cravador, A., Steinert, M., and Hamers, R. (1983) Gene activation and re-expression of a *Trypanosoma brucei* variant surface glycoprotein. *EMBO J.*, **2**, 1185.

169. Parsons, M., Nelson, R. G., Stuart, K., and Agabian, N. (1984) Variant antigen genes of *Trypanosoma brucei*: genomic alteration of a spliced leader orphon and retention of expression-linked copies during differentiation. *Proc. Natl. Acad. Sci. USA*, **81**, 684.

170. Delauw, M. F., Pays, E., Steinert, M., Aerts, D., Van Meirvenne, N., and Le Ray, D. (1985) Inactivation and reactivation of a variant-specific antigen gene in cyclically transmitted *Trypanosoma brucei*. *EMBO J.*, **4**, 989.

171. Cornelissen, A. W. C. A., Johnson, P., Van der Ploeg, L. H. T., and Borst, P. (1985) Two simultaneously active VSG gene transcription units in a single *Trypanosoma brucei* variant. *Cell*, **41**, 825.

172. Shea, C., Glass, D. J., Parangi, S., and Van der Ploeg, L. H. T. (1986) Variant surface glycoprotein (VSG) gene expression sites switches in *Trypanosoma brucei*. *J. Biol. Chem.*, **261**, 5056.

173. Bernards, A., Van Harten-Loosbroek, N., and Borst, P. (1984) Modification of telomeric DNA in *Trypanosoma brucei*, a role in antigenic variation? *Nucleic Acids Res.*, **12**, 4153.

174. Pays, E., Delauw, M. F., and Steinert, M. (1984) Possible DNA modification in GC dinucleotides of *Trypanosoma brucei* telomeric sequences; relationship with antigen gene transcription. *Nucleic Acids Res.*, **12**, 5235.

175. Pays, E., Laurent, M., Delinte, K., Van Meirvenne, N., and Steinert, M. (1983) Differential size variations between transcriptionally active and inactive telomeres of *Trypanosoma b. brucei*. *Nucleic Acids Res.*, **11**, 8137.

176. Myler, P. J., Aline, R. F., Scholler, J. K., and Stuart, K. D. (1988) Changes in telomere length associated with antigenic variation in *Trypanosoma brucei*. *Mol. Biochem. Parasitol.*, **29**, 243.

177. Baltz, T., Giroud, C., Baltz, D., Roth, C., Raibaud, A., and Eisen, H. (1986) Stable expression of two variable surface glycoproteins by cloned *Trypanosoma equiperdum*. *Nature*, **329**, 602.

178. Esser, K. M. and Schoenbechler, M. J. (1985) Expression of two variant surface glycoproteins on individual African trypanosomes during antigen switching. *Science*, **229**, 190.

179. Kosinski, R. J. (1980) Antigenic variation in trypanosomes: a computer analysis of variant order. *Parasitology*, **80**, 343.

180. Abiri, D., Van der Ploeg, L. H. T., and Agur, Z. (1989) The ordered appearance of antigenic variants of African trypanosomes explained in a mathematical model based on a stochastic switch process and immune-selection against putative switch intermediates. *Proc. Natl. Acad. Sci. USA*, in press.

181. Eid, J. and Sollner-Webb, B. (1987) Efficient introduction of plasmid DNA into *Trypanosoma brucei* and transcription of a transfected chimeric gene. *Proc. Natl. Acad. Sci. USA*, **84**, 7812.

182. Gibson, W. C., White, T. C., Laird, P. W., and Borst, P. (1987) Stable introduction of exogenous DNA into *Trypanosoma brucei*. *EMBO J.*, **6**, 2457.

183. Van der Ploeg, L. H. T., Liu, A. Y. C., Michels, P. A. M., De Lange, T., Borst, P., Majumder, H. K., Weber, H., Veeneman, G. H., and Van Boom, J. H. (1982) RNA splicing is required to make the messenger RNA for a variant surface antigen in trypanosomes. *Nucleic Acids Res.*, **10**, 3591.

184. Boothroyd, J. C. and Cross, G. A. M. (1982) Transcripts for different variant surface glycoproteins of *Trypanosoma brucei* have a short, identical exon at their 5' end. *Gene*, **20**, 281.

185. Guyaux, M., Cornelissen, A. W. C. A., Pay, E., Steinert, M., and Borst, P. (1985) *Trypanosoma brucei*: a surface antigen mRNA is discontinuously transcribed from two distinct chromosomes. *EMBO J.*, **4**, 995.

186. Freistadt, M. S., Cross, G. A. M., Branch, A. D., and Robertson, H. D. (1987) Direct analysis of the mini-exon donor RNA of *Trypanosoma brucei*: detection of a novel cap structure also present in messenger RNA. *Nucleic Acids Res.*, **15**, 9861.

187. Perry, K. L., Watkins, K. P., and Agabian, N. (1987) Trypanosome mRNAs have unusual 'cap 4' structures acquired by addition of a splice leader. *Proc. Natl. Acad. Sci. USA*, **84**, 8190.

188. Sutton, R. E. and Boothroyd, J. C. (1988) The cap of both mini-exon-derived RNA and mRNA of trypanosomes is 7-methylguanosine. *Mol. Cell. Biol.*, **8**, 494.

189. Laird, P. W., Kooter, J. M., Loosbroek, N., and Borst, P. (1985) Mature mRNAs of *Trypanosoma brucei* possess a 5' cap acquired by discontinuous RNA synthesis. *Nucleic Acids Res.*, **13**, 4253.

190. De Lange, T., Michels, P. A. M., Veerman, H. J. G., Cornelissen, A. W. C. A., and Borst, P. (1984) Many trypanosome messenger RNAs share a common 5' terminal sequence. *Nucleic Acids Res.*, **12**, 3777.

191. Sather, S. and Agabian, N. (1985) A 5' spliced leader is added in *trans* to both alpha- and beta-tubulin transcripts in *Trypanosoma brucei Proc. Natl. Acad. Sci. USA*, **82**, 5695.

192. Walder, J. A., Eder, P. S., Engman, D. M., Brentano, S. T., Walder, R. Y., Knutzon, D. S., Dorfman, D. M., and Donelson, J. E. (1986) The 35-nucleotide splice leader sequence is common to all trypanosome messenger RNAs. *Science*, **233**, 569.

193. Cornelissen, A. W. C. A., Verspieren, M. P., Toulme, J. J., Swinkels, B. W., and Borst, P. (1986) The common 5' terminal sequence on trypanosome mRNAs: a target for anti-messenger oligodeoxynucleotides. *Nucleic Acid Res.*, **14**, 5605.

194. De Lange, T., Liu, A. Y. C., Van der Ploeg, L. H. T., Borst, P., Tromp, M. C., and Van Boom, J. H. (1983) Tandem repetition of the 5' mini-exon of variant surface glycoprotein genes; a multiple promoter for VSG gene transcription. *Cell*, **34**, 891.

195. Nelson, R. G, Parsons, M., Barr, P., Stuart, K., Selkirk, M., and Agabian, N. (1983) Sequences homologous to the variant antigen mRNA spliced leader are located in tandem repeats an variable orphons in *Trypanosoma brucei*. *Cell*, **34**, 901.

196. Nelson, R. G., Parsons, M., Selkirk, M., Newport, G., Barr, P., and Agabian, N. (1984) Sequences homologous to variant antigen mRNA spliced leader in trypanosomatidae which do not undergo antigenic variation. *Nature*, **308**, 665.

197. Kooter, J. M., De Lange, T., and Borst, P. (1984) Discontinuous synthesis of mRNA in trypanosomes. *EMBO J.*, **3**, 2387.

198. Parsons, M., Nelson, R. G., Watkins, K. P., and Agabian, N. (1984) Trypanosome mRNAs share a common 5' spliced leader sequence. *Cell*, **38**, 109.

199. Campbell, D. A., Thornton, D. A., and Boothroyd, J. C. (1984) Apparent discontinuous transcription of *Trypanosoma brucei* variant surface antigen genes. *Nature*, **311**. 350.

200. Freistadt, M. (1988) Detection of a possible *trans*-splicing intermediate in *Trypanosoma brucei*. *Nucleic Acids Res.*, **16**, 7720.

201. Lee, M. G. -S. and Van der Ploeg, L. H. T. (1989) Transcription of the heat shock 70 locus in *Trypanosoma brucei*. Submitted.

202. Kooter, J. M. and Borst, P. (1984) Alpha-amanintin insensitive transcription of variant surface glycoprotein genes provides further evidence for discontinuous transcription in trypanosomes. *Nucleic Acids Res.*, **12**, 9457.

203. Laird, P. W., Zomerdijk, J. C. B. M., De Korte, D., and Borst, P. (1987) *In vitro* labelling of intermediates in the discontinuous synthesis of mRNAs in *Trypanosoma brucei*. *EMBO J.*, **6**, 1055.

204. Murphy, W. J., Watkins, K. P., and Agabian, N. (1986) Identification of a novel Y branch structure as an intermediate of trypanosome mRNA processing: evidence for *trans* splicing. *Cell*, **47**, 517.

205. Sutton, R. E. and Boothroyd, J. C. (1986) Evidence for *trans* splicing in trypanosomes. *Cell*, **47**, 527.

206. Ralph, D., Huang, J. and Van der Ploeg, L. H. T. (1988) Physical identification of branched intron side-products of splicing in *Trypanosoma brucei*. *EMBO J.*, **8**, 2539.

207. Sutton, R. E. and Boothroyd, J. C. (1988) Trypanosome *trans*-splicing utilizes 2′–5′ branches and a corresponding debranching activity. *EMBO J.*, **7**, 1431.

208. Tschudi, C., Richards, F. F., and Ullu, E. (1986) The U2 RNA analogue of *Trypanosoma brucei gambiense*: implications for a splicing mechanism in trypanosomes. *Nucleic Acids Res.*, **14**, 8893.

209. Krause, M. and Hirsh, D. (1987) A *trans*-spliced leader sequence on actin mRNA in *C. elegans*. *Cell*, **49**, 753.

210. Van Doren, K. and Hirsh, D. (1988) *Trans*-spliced leader RNA exists as small ribonucleoprotein particles in *Caenorhabditis elegans*. *Nature*, **335**, 556.

211. Miller, A. I. and Wirth, D. F. (1988) *Trans*-splicing in *Leishmania enriettii* and identification of ribonucleoprotein complexes containing the spliced leader and U2 equivalent RNAs. *Mol. Cell. Biol.*, **8**, 2597.

212. Bruzik, J. P., Van Doren, K., Hirsh, D., and Steitz, J. A. (1988) *Trans* splicing involves a novel form of small nuclear ribonucleoprotein particles. *Nature*, **335**, 559.

213. Gonzalez, A., Lerner, T. J., Huecas, M., Sosa-Pineda, B., Nogueira, N., and Lizardi, P. M. (1985) Apparent generation of a segmented mRNA form two separate tandem gene families in *Trypanosoma cruzi*. *Nucleic Acids Res.*, **13**, 5789.

214. Tschudi, C. and Ullu, E. (1988) Polygene transcripts are precursors to calmodulin mRNAs in trypanosomes. *EMBO J.*, **7**, 455.

215. Glass, D., Polvere, R. I., and Van der Ploeg, L. H. T. (1986) Conserved sequences and transcription of the hsp 70 gene family in *Trypanosoma brucei*. *Mol. Cell. Biol.*, **6**, 4657.

216. Lee, M. G. -S., Atkinson, B. L., Giannini, S. H., and Van der Ploeg, L. H. T. (1989) Structure and expression of the hsp 70 family of *Leishmania major*. *Nucleic Acids Res.*, **16**, 9567.

217. Thomashow, L. S., Milhausen, M., Rutter, W. J., and Agabian, N. (1983) Tubulin genes are linked and clustered in the genome of *Trypanosoma brucei*. *Cell*, **32**, 35.

218. Imboden, M. A., Laird, P. W., Affolter, M., and Seebeck, T. (1987) Transcription of the intergenic regions of the tubulin gene clusters of *Trypanosoma brucei*: evidence for a polycistronic transcription unit in a eukaryote. *Nucleic Acids Res.*, **15**, 7357.

219. Gibson, W., Swinkels, B. W., and Borst, P. (1988) Post-transcriptional control of the differential expression of phosphoglycerate kinase genes in *Trypanosoma brucei*. *J. Mol. Biol.*, **201**, 315.

220. Muhich, M. L. and Boothroyd, J. C. (1988) Polycistronic transcripts in trypanosomes and their accumulation during heat shock: evidence for a role in mRNA synthesis. *Mol. Cell. Biol.*, **8**, 3837.

221. White, T. and Borst, P. (1987) RNA end labeling and RNA ligase activities can produce a circular rRNA in whole cell extracts from trypanosomes. *Nucleic Acids Res.*, **15**, 3275.

222. Laird, P. W., ten Asbroek, A. L. M. A., and Borst, P. (1987) Controlled turnover and 3′ trimming of the *trans* splicing precursor of *Trypanosoma brucei*. *Nucleic Acids Res.*, **15**, 10087.

223. Huang, J. and Van der Ploeg, L. H. T. (1988) A 5′ exo-ribonuclease and RNA ligase in T. brucei. Nucleic Acids Res., **16**, 9737.

224. Sentenac, A. (1985) Eukaryotic RNA polymerases. *CRC Crit. Rev. Biochem.*, **18**, 31.

225. Cornelissen, A. W. C. A., Evers, R., and Kock, J. (1988) Structure and sequence of genes encoding subunits of eukaryotic RNA polymerases. *Oxford Surveys Euk. Gen.*, **5**, 91.

226. Cornelissen, A. W. C. A., Evers, R., Grondal, E., Hammer, A., Jess, W., and Kock, J. (1989) Transcription and RNA polymerases in *Trypanosoma brucei*. *Nova Acta Leopoldina*, in press.

227. Kitchen, P. A., Ryley, J. F., and Gutteridge, W. (1984) The presence of unique DNA-dependent RNA polymerases in trypanosomes. *Comp. Biochem. Physiol.*, **77B**, 223.

228. Earnshaw, L., Beebee, T. J. C., and Gutteridge, W. (1987) Demonstration of RNA polymerase multiplicity in *Trypanosoma brucei*. *Biochem. J.*, **241**, 649.

229. Evers, R., Hammer, A., Kock, J., Jess, W., Borst, P., Memet, S., and Cornelissen, A. W. C. A. (1989) *Trypanosoma brucei* contains two RNA polymerase II largest subunit

genes with an altered C-terminal domain. *Cell,* **56**, 585.

230. Kock, J., Evers, R., and Cornelissen, A. W. C. A. (1988) Structure and sequence of the gene for the largest subunit of trypanosomal RNA polymerase III. *Nucleic Acids Res.,* **16**, 8753.

231. Coulter, D. E. and Greenleaf, A. L. (1982) Properties of mutationally altered RNA polymerase II of *Drosophila. J. Biol. Chem.,* **257**, 1945.

232. Coulter, D. E. (1986) A mutation in the largest subunit of RNA polymerase II alters chain elongation *in vitro. J. Biol. Chem.,* **260**, 13190.

233. Brodner, O. G. and Wieland, Th. (1976) Identification of the amatoxin-binding subunit of RNA polymerase B by affinity labeling experiments. Subunit B3 – the true amatoxin receptor protein of multiple RNA polymerase B. *Biochemistry,* **15**, 3480.

234. White, T., Rudenko, G., and Borst, P. (1986) Three small RNAs within the 10 kb trypanosome rRNA transcription unit are analogous to domain VII of other eukaryotic 28S rRNAs. *Nucleic Acids Res.,* **14**, 9471.

235. Ehlers, B., Czichos, J., and Overath, P. (1987) RNA turnover in *Trypanosoma brucei. Mol. Cell. Biol.,* **7**, 1242.

236. Overath, P., Czichos, J., Stock, U., and Nonnengaesser, C. (1983) Repression of glycoprotin synthesis and release of surface coat during transformation of *Trypanosoma brucei. EMBO J.,* **2**, 1721.

237. Van der Ploeg, L. H. T., Giannini, S. H., and Cantor, C. R. (1985) Heat-shock genes: regulatory role for differentiation in parasitic protozoa. *Science,* **228**, 1443.

238. Rudenko, G., Bishop, D., Gottesdiener, K., and Van der Ploeg, L. H. T. (1989) Alpha-aminitin resistant transcription of protein coding genes in insect and blood stream forms of *Trypanosoma brucei. Embo J.,* in press.

239. Roditi, I., Carrington, M., and Turner, M. J. (1987) Expression of a polypeptide containing a dipeptide repeat is confined to the insect stage of *T.brucei. Nature,* **325**, 272.

240. Mowatt, M. R .and Clayton, C. (1987) Developmental regulation of a novel repetitive protein of *Trypanosoma brucei. Mol. Cell. Biol.,* **7**, 2838.

241. Mowatt, M. R. and Clayton, C. (1988) Polymorphism in the procyclic acidic repetitive protein gene family of *Trypanosoma brucei. Mol. Cell. Biol.,* **8**, 4055.

242. Clayton, C. and Mowatt, M. R. (1989) The procyclic acidic repetitive proteins of *Trypanosoma brucei:* purification and evidence for a glycosyl-phosphatidyl inositol membrane anchor. *J. Biol. Chem.,* **264**, 15088.

243. Vickerman, K. (1985) Development cycles and biology of phatogenic trypanosomes. *Br. Med. Bull.,* **41**, 105.

244. Pearson, T. W., Moloo, S. K., and Jenni, L. (1987) Culture form and tsetse fly midgut form procyclic *Trypanosoma brucei* express common proteins. *Mol. Biochem. Parasitol.,* **25**, 273.

245. Aksoy, S., Lalor, T. M., Martin, J., Van der Ploeg, L. H. T., and Richards, F. F. (1987) Multiple copies of a rctroposon interrupt spliced-leader RNA genes in the African trypanosome, *T. gambiense. EMBO J.,* **12**, 3819.

246. Hajduk, S. L., Cameron, C. R., Barry, J. D., and Vickerman, K. (1981) Antigenic variation in cyclically transmitted *Trypanosoma brucei* variable antigen type composition of metacyclic trypanosome populations from the salivary glands of *Glossina morsitans. Parasitology,* **83**, 595.

247. Hajduk, S. L. and Vickerman, K. (1981) Antigenic variation in cyclically transmitted *Trypanosoma brucei.* Variable antigen composition of the first parasitaemia in mice bitten by trypanosome-infected *Glossina morsitans. Parasitology,* **83**, 609.

248. Barry, J. D., Crowe, J. S., and Vickerman, K. (1985) Neutralization of individual variable antigen types in metacyclic populations of *Trypanosoma brucei* does not prevent their subsequent expression in mice. *Parasitology,* **90**, 79.

249. Cornelissen, A. W. C. A., Bakkern, G. A. M., Barry, J. D., Michels, P. A. M., and Borst, P. (1985) Characteristics of trypanosome variant antigen genes active in the tsetse fly. *Nucleic Acids Res.,* **13**, 4661.

250. Esser, K. M., Schoenbechler, M. J., and Gingrich, J. B. (1982) *Trypanosoma rhodesiense* blood forms express all antigen specificities to protection against metacyclic (insect-form) challenge. *J. Immunol.,* **129**, 1715.

251. Lenardo, M. J., Esser, K. M., Moon, A. M., Van der Ploeg, L. H. T., and Donelson, J. E. (1986) Metacyclic variant surface glycoprotein genes of *Trypanosoma brucei* subsp.

rhodesiense are activated *in situ*, and their expression is transcriptionally regulated. *Mol. Cell. Biol.,* **6**, 1991.

252. Delauw, M. F., Laurent, M., Paindavoine, P., Aerts, D., Pays, E., Le Ray, D., and Steinert, M. (1987) Characterization of genes coding for two major metacyclic surface antigens in *Trypanosoma brucei*. *Mol. Biochem. Parasitol.,* **23**, 9.

253. Frasch, A. C. C., Borst, P., and Van den Burg, J. (1982) Rapid evolution of genes coding for variant surface glycoproteins in trypanosomes. *Gene,* **17**, 197.

254. Michels, P. A. M. (1986) Evolutionary aspects of trypanosomes: analysis of genes. *J. Mol. Evol.,* **24**, 45.

255. Bernards, A., Van der Ploeg, L. H. T., Gibson, W. C., Leegwater, P., Eijgenraam, F., De Lange, T., Weijers, P., Calafat, J., and Borst, P. (1986) Rapid change of the repertoire of variant surface glycoprotein genes in trypanosomes by gene duplication and deletion. *J. Mol. Biol.,* **190**, 1.

256. Hasan, G., Turner, M. J., and Cordingly, J. S. (1984) Complete nucleotide sequence of an unusual mobile element from *Trypanosoma brucei*. *Cell,* **37**, 333.

257. Kimmel, B. E., Ole-Moiyoi, O. K., and Young, J. R. (1987) Ingi, a 5.2 kb dispersed sequence element from *Trypanosoma brucei* that carries half of a smaller mobile element at either end and has homology with the mammalian LINEs. *Mol. Cell. Biol.,* **7**, 1465.

258. Pays, E. and Steinert, M. (1984) Telomeric DNA rearrangements and antigenic variation in African trypanosomes. In *Hormones and Cell Regulation*. INSERM European Symposium, Vol. 8. Dumont, J. E. and Nunez, J. (eds), Elsevier Science Publishers, Amsterdam, p. 289.

259. Rice-Ficht, A. C., Chen, K. K., and Donelson, J. E. (1982) Point mutations during generation of expression-linked extra copy of trypanosome surface glycoprotein gene. *Nature,* **298**, 676.

260. Bellofatto, V. and Cross, G. (1989) Expression of a bacterial gene in trypanosomatic protozoan. *Science,* **244**, 1167.

261. Mottram, J., Perry, K. L., Lizzardi, P. M., Luhrmann, R., Agabian, N., and Nelson, R. G. (1989) Isolation and sequence of four small nuclear U RNA genes of *Trypanosoma brucei* subsp. *brucei*. *Mol. Cell. Biol.,* **9**, 1212.

262. Smith, J. L., Levin, J. R., Ingles, J. C., and Agabian, N. (1989) In trypanosomes the homolog of the largest subunit of RNA polymerase II is encoded by two genes and had a highly unusual C-terminal domain structure. *Cell,* **56**, 815.

263. Rudenko, G. and Van der Ploeg, L. H. T. (1989) Transcription of telomere repeats in protozoa. *EMBO J.,* **8**, 2633.

3

DNA amplification in eukaryotes

George R.Stark, Michelle Debatisse,
Geoffrey M.Wahl, and David M.Glover

1. Introduction

The phenomenon of DNA (or gene) amplification is widespread in biology, with examples throughout the phylogenetic tree. Amplification is defined as an increase in the *relative* amount of a gene or DNA sequence within a cell, usually involving less DNA than the amount contained in a single chromosome. Thus we mean to exclude phenomena such as polyteny, in which all or most of the genome is increased in copy number within each cell, and hyperploidy, in which additional copies of one or more whole chromosomes are present due to unequal segregation at mitosis. Amplification is well known as a normal phenomenon in lower eukaryotes, where developmentally programmed increases in copy number lead to rapid gene expression in particular cells or tissues at specific times, when the need for a particular gene product cannot be met by transcription and translation working at their maximal efficiencies. It is perhaps more usual for an organism to meet such demands by having multiple copies of the genes within its haploid genetic complement. Nevertheless, several examples of developmentally regulated gene amplification are known in both unicellular and multicellular eukaryotic organisms. In Section 3 we concentrate upon the phenomenon as it occurs in dipteran flies, which provide the best studied cases.

In a developmentally programmed amplification, large numbers of cells carry out the event at the same time, so that it is readily amenable to study. Not surprisingly, therefore, we know quite a lot about the mechanisms involved. Amplification is also well known as an abnormal phenomenon which is not developmentally programmed. Examples are resistance to toxic agents in lower eukaryotic organisms, including insects, and drug resistance and amplification of oncogenes in mammalian cells. Abnormal amplification happens rarely (for example 10^{-5} is a commonly observed frequency for amplification of a target gene in a population of mammalian cells), and so it has not been possible to obtain relatively homogeneous populations of cells that are all amplifying the same locus at about the same time. Consequently, we know less about abnormal amplification and, despite intensive effort, we cannot yet be certain of the mechanism in a single case. Nevertheless, good progress has been made recently and the pace is accelerating.

It is important to understand how over-accumulation of particular gene products, derived from amplified genes, contribute to particular phenomena such as metastasis. Quite separately, it is also important to know about the mechanisms governing amplification events and to discover which functions (gene products) are responsible, both for programmed amplifications and also for abnormal amplifications, which can often be stimulated transiently by external events or more stably by heritable changes within cells. It seems likely that unprogrammed amplification is but one manifestation of more fundamental defects that lead to chromosomal abnormalities such as deletion, inversion, translocation, and possibly even chromosome loss (see Section 7.3). Perhaps by studying mechanisms of unprogrammed amplification we shall be led to genes, the abnormal expression of which is responsible for underlying defects in the regulation of chromosomal integrity.

Another consequence of studying gene amplification in cultured cells has been its biotechnological application. Not only the gene selected by the drug but also flanking sequences are amplified. Thus, by linking a gene that encodes a protein of commercial importance to one whose amplification can be selected with a drug, it is possible to amplify both genes and express both gene products at a high level.

The subject of amplification has been reviewed quite regularly since 1982. The interested reader is therefore referred to the relatively general reviews listed below for historical perspective and also for greater detail in some subjects. For example, the 1982 review of Cowell (1) gives excellent background information on the cytologic manifestations of amplification in mammalian cells, and the 1982 book edited by Schimke (2), reporting the proceedings of an early conference, is full of interesting facts and hypotheses that are still quite relevant. There are several more recent general reviews (3–14). Additional reviews covering more specific aspects are referred to below in the appropriate sections.

2. Developmentally regulated rDNA amplification

2.1 rDNA amplification in ciliates

An example of gene amplification in a unicellular organism is provided by the rDNA in the macronuclei of *Tetrahymena*. Ciliated protozoa have two nuclei, a diploid micronucleus that serves as the germ line, and a polyploid macronucleus that is transcriptionally active. The macronucleus is generated from the newly formed zygotic micronucleus after conjugation, a process which is accompanied by chromosome breakage, elimination of some sequences and acquisition of telomore-like sequences by those chromosomal fragments required for vegetative growth (15). The rDNA represents at least one gene set that is amplified by about 200-fold during this process. The single 11 kb rDNA unit is excised from the micronuclear DNA and converted into an extrachromosomal palindromic dimer of 21 kb, which is subsequently replicated. Sequences that may play a role in the excision and dimerization of the rDNA unit have been identified adjacent to the rDNA in its chromosomal location in the micronucleus (16). It is likely that other *Tetrahymena* genes may undergo amplification by a similar process.

2.2 rDNA amplification in amphibian oocytes

In the development of amphibians such as *Xenopus laevis*, a high level of replication of the rDNA cistrons occurs during oogenesis to provide the mature oocyte with about 10^{12} ribosomes. There is an initial amplification of the rDNA by 10- to 40-fold in the gonial cells, which is followed by a subsequent 2500-fold increase. The latter burst of amplification occurs over a period of about 3 weeks in early meiotic prophase. Pulse-labeled autoradiographic studies carried out in conjunction with electron microscopy favor a rolling circle model of replication in which the amplifying DNA becomes extrachromosomal (17). As the spacers between the rDNA genes are highly polymorphic, different rDNA units can be readily distinguished by restriction endonuclease cleavage. Comparison of the cleavage pattern of somatic versus amplified rDNA indicates that only a subset of the rDNA undergoes amplification, consistent with the extrachromosomal replication model (18).

The most detailed studies of developmentally regulated gene amplification have been carried out on insects, especially *Drosophila melanogaster*. We will concentrate upon these studies in the next section.

3. Gene amplification during the development of dipteran flies

3.1 Cell division and polytenization during dipteran development

The ability to over-replicate certain regions of the genome in flies such as *Drosophila* is intimately connected with the modes of chromosomal replication, which themselves reflect variations in patterns of the cell-cycle in development of the organism. The *Drosophila* embryo is initially a syncytium for about the first 2 h of its development. During this time the number of nuclei increase from the single zygotic nucleus of the fertilized embryo to approximately 5000 nuclei at the time of cellularization (19). Before cellularization occurs, there are 13 rapid nuclear division at 10 min intervals. These divisions occur initially while the nuclei are in the interior of the embryo, but at nuclear cycle 9 the majority of nuclei migrate to the cortex and continue their division cycles (20). The yolk nuclei remain in the interior of the embryo, lose their centrosomes, and cease dividing, but DNA replication continues and these nuclei become polyploid. This is the first example during development of a choice between a cell remaining diploid and continuing to undergo proliferation, or becoming polyploid (see Section 3.3).

Following cellularization, the cell-cycle lengthens, allowing the major transcriptional activity of zygotic nuclei to begin. Three to four rounds of subsequent mitosis occur within distinct domains of the embryo and are, in part, responsible for morphogenesis. The cell-cycle time of these divisions is between 1 and 2 h (21). Most larval development then comprises cell growth and polyploidization in the absence of cell division. The exceptions are the imaginal tissues and cells of the nervous system. These continue to proliferate, eventually forming adult structures follow-ing pupariation. These cells have a slower cycle time, ranging from 12 to 48 h. Many cell-cycle mutants die because the imaginal tissues cannot proliferate to form adult structures (22). The organisms can survive until this developmental stage, however, because their heterozygous mothers have produced enough wild-type gene product for embryonic development, following which there is a reduced requirement for these proteins in the period of larval growth with its extensive polyploidization.

Another group of mitotic mutations have a maternal effect lethal phenotype. These correspond to genes that direct synthesis of proteins in the developing oocyte that are required for the rapid cleavage divisions in the first 2 h of embryogenesis. A female, homozygous for a mutation in such a gene, lays an egg in which a protein essential for mitosis is either missing or defective (see ref. 23 for a review). The mutation, *gnu*, is one such example. Females, homozygous for this mutation, lay eggs in which there is no nuclear division. DNA synthesis continues, however, such that a small number of giant nuclei are produced in the embryo (24,25). It is now known that fertilization is not required for the eggs of the homozygous *gnu* mothers to develop giant nuclei. This has led to the idea that the *gnu* mutation overcomes the block to DNA synthesis that is normally imposed upon the four products of female meiosis, the female pronucleus and the three polar bodies. In some way this uncontrolled DNA replication uncouples the giant nuclei from the mitotis cycle, leading to the dissociation of centrosomes and their autonomous replication within the cytoplasm of these embryos. One might also imagine that the *gnu* phenotype reflects a defect in the mechanism controlling the switch between diploidy and polyploidy in *Drosophila*.

The control of diploid cell proliferation is beyond the scope of this chapter, although some discussion of the developmental transitions between the diploid and polytene states is necessary in considering the phenomenon of gene amplification, which in dipterans appears to take place only in polyploid tissues. Polyploidization is initiated in several tissues during mid-embryogenesis. By the time that the larvae hatch, the chromosomes of the salivary gland cells, for example, have undergone three cycles of endoreduplication (replication of the chromatids without mitotic segregation). In the late third instar larvae, prior to pupation, these same cells have completed nine or ten rounds of endoreduplication and usually have ploidies of greater than 1000-fold. Many other tissues become polyploid during either larval, pupal, or adult development. In *D. melanogaster*, the bristle mother cells within the pupal epidermis become polyploid during pupal development, and the nurse and follicle cells of the adult ovary become polyploid during oogenesis. The follicle cells surround the developing egg chamber and undergo about five rounds of endoreduplication. Some chromosomal regions undergo additional rounds of DNA replication resulting in their amplification to higher levels than surrounding chromosomal sequences (see Section 3.4). The nurse cells form part of the egg chamber and reach ploidies equivalent to those of the salivary gland cells. Their chromosomes are not condensed, however, and many genes are active, providing the developing oocyte with the large number of maternal gene products needed for early embryogenesis. These cells have essentially amplified the major proportion of their genomes to meet the exacting requirements for gene expression during the rapid development of the insect egg.

3.2 The distribution of replication origins in *Drosophila* chromosomes

In order that the nuclei division cycle of cleavage embryos is accomplished within a 10 min interval, there is an extraordinarily rapid period of DNA synthesis in which the genome is replicated with $3-4$ min. This is achieved using as many as 20 000 bidirectional replicational origins which can be visualized directly by electron microscopy (26,27) (*Figure 1*). These replication forks have the typical structure expected of replicative intermediates, with short single-stranded regions on one strand of each fork. The average spacing between these replicons is 7.9 kb and spacings larger than about 90 kb are virtually never seen. The elongation rate is calculated to be 2.6 kb min^{-1} and would be sufficient to replicate 19 kb between two replication origins within an S-phase of $3-4$ min. The reinitiation of replication within a replicon was never observed in over 1000 replication intermediates that were examined.

Following cellularization, the length of S-phase increases to about 20 min. A similar number of replicons can be observed in these nuclei however, and the average spacing is 10.6 kb (28). There is considerable asynchrony in the initiation of replication which probably accounts for the increased length of S-phase at this time of development. The pattern of DNA replication in cells derived from larval brains and imaginal disks,

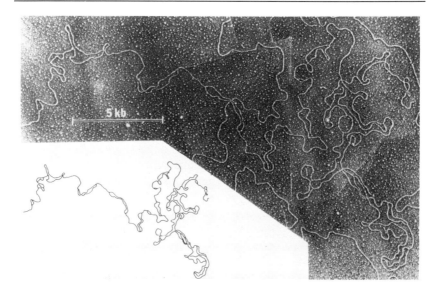

Figure 1. An electron micrograph of replicating chromosomal DNA from *D.melanogaster* cleavage embryos (taken from ref. 26). This particular segment of DNA is 119 kb long and contains 23 bidirectionally replicating loops. In contrast to DNA from amplifying regions (see *Figure 3*), replicating loops have never been observed within loops, Thus, in mitotically active diploid tissues, there exists a mechanism to prevent reinitiation of chromosomal replication before one round has been completed.

or tissue culture cell lines is very different (27,29). In these cases replicative forms cannot be observed in the electron microscope and it has been necessary to follow replication by autoradiography following pulse-labeling with [^{3}H]thymidine. These studies show that fewer forks are used during the 12 h S-phases of these cells, and that the average replicon is between 40 and 100 kb in length, although the rate of fork elongation has been calculated to be equivalent to that in the cleavage embryo.

Thymidine incorporation has also been used to study S-phases in polytene cells (e.g. 30,31). These and other data suggest that DNA replication does not occur continuously within these cells but rather in a cycle, in which there are three characteristic patterns of [^{3}H]thymidine incorporation.

(i)   An early pattern in which only some chromosomal regions are labeled and in which the chromocenter is unlabeled.

(ii)   An intermediate stage in which there is continuous labeling along the entire lengths of the chromosomes and the chromocenter.

(iii) A later form of replication in which only the chromocenter and a few bands show labeling.

The pattern of late replication would seem to be site-specific since when these regions are translocated to different sites, they display the same temporal pattern of DNA replication as in wild-type chromosomes.

3.3 Differential replication of DNA in polytene chromosomes

The differential replication of different chromosomal regions is characteristic of polyploid cells in insects. Direct examination of condensed chromosomes of diploid cells and of polytene chromosomes of salivary gland cells in the light microscope indicates that the fraction of the chromosome comprising centromeric heterochromatin is greatly reduced in the polytene state as compared with the diploid state. The heterochromatic regions that surround the centromeres of the autosomes and the X chromosome are comprised mainly of satellite DNA sequences. These, together with all of the Y chromosome, are significantly under-replicated in the polytene nuclei of the salivary gland cells. The satellite DNA sequences are also under-represented in many other polytene cells of *D.melanogaster*, including the nurse cells (30 – 32). This under-replication of satellite sequences may not occur in all tissues and there are variations in the degree of under-replication in the same tissue of different species of dipterans. The pupal ovaries of *D.virilus*, for example, have 30% more satellite DNA than do larval brains (33), although satellite sequences are under-replicated in the adult ovary (34). In *Calliphora erythrocephela* (blowfly), satellite sequences replicate to an equivalent extent as euchromatic sequences in the nurse cells, whereas in other polytene tissues, satellite DNAs are under-replicated (35). In general, however, it seems that satellite sequences are under-replicated in polytenization. This could result either from satellite DNA undergoing fewer complete rounds of chromosome replication relative to euchromatic sequences, or from incomplete chromosomal replication in each S-phase reflecting the late replication of satellite sequences within the polytenization cycle (Section 3.2).

3.3.1 rDNA

rDNA sequences are also differentially replicated during the polytenization process. This has been reviewed by Beckingham (36) and only the salient features of the phenomenon will be addressed here. In *D.melanogaster* salivary glands, for example, there is only ¼ to ⅛ of the expected number of rDNA genes (37,38). The under-replication of rDNA is also found in many other polytene tissues.

A high proportion of the rDNA genes of *D.melanogaster* contains one of two types of insertion (39). These sequences, together with the highly polymorphic spacers that separate the rDNA transcription units, allow many different classes of rDNA to be recognized in Southern blotting experiments. This type of analysis has provided an explanation for the observation of Spear and Gall (37) that there are equal amounts of rDNA in X – X and X – 0 flies. It appears that only one of the two nucleolus organizers (NOs) becomes polytenized in any given fly (40,41), the NOs of certain chromosomes being dominant over others in their ability to undergo replication (42,43).

There is some indication that rDNA genes with inserts, which are unable to produce rRNA, are selectively under-replicated. As genes with inserts are randomly intermingled with active genes in *D.melanogaster*, this differential replication is not as marked as in species in which these two types of rDNA tend to occur in blocks. The under-replication of clusters of rDNA genes with inserts is found in both *D.hydei* and *C.erythrocephela* (reviewed in ref. 36). As with satellite sequences, rDNA is more extensively replicated during the polytenization of nurse cells, enabling these cells to produce the large numbers of ribosomes required by the egg. Nevertheless, there still appears to be selective replication. In *C.erythrocephela*, for example, active rDNA is fully replicated, whereas rDNA genes with inserts are under-replicated 8-fold (44). In general it seems that nurse cells contain between four and eight times more functional rDNA than cells of similar ploidy from other tissues.

3.3.2 Euchromatic sequences

Many segments of the *Drosophila* genome have now been cloned and so it is possible to compare the representation of these sequences in polytene and diploid tissues. Furthermore, it is of interest to know whether the DNA in the banded regions of salivary gland chromosomes differs in its degree of replication from the DNA in the interband regions. Spierer and Spierer (45) have analyzed 315 kb of continuous sequence between the salivary gland chromosome regions 87D5 to 87E5,6. They found no significant differences in the relative amounts of DNA from the diploid and polytene cells in this region, irrespective of the distribution of bands and interbands. The conclusion that there is no difference in copy number between euchromatic genes in diploid or polytene tissues has been confirmed in several studies with cloned genes (e.g. 46). However, certain regions of the polytene chromosomes of *Drosophila* have constrictions which have long been thought to represent under-replicated regions. The DNA content of one of these constrictions around the *Bithorax* locus at 89E has been examined and found to be significantly under-replicated (45). Similarly, the tandemly arranged histone genes, present within a constriction at 39DE, are under-represented by about 2-fold in salivary gland DNA (40).

Conversely, there are examples of euchromatic sequences at certain chromosomal regions that are over-replicated, resulting in the amplification of specific genes at particular developmental stages. This occurs in the DNA puffs of certain sciarid flies and in the chorion gene clusters of *D.melanogaster*. Discussion of these phenomena forms the basis of the next section.

3.4 Amplification of single copy genes

3.4.1 DNA puffs

The first direct observation of developmentally regulated amplification of regions of insect chromosomes was made in the mid-1950s by Breuer and Pavan (47) involving the salivary gland chromosomes of the larvae of the *Rhynchosciara angelae* (americana). Several specific regions (the DNA puffs) of these polytene chromosomes were observed to increase in size and staining intensity before opening into large transcriptionally active puffs. Subsequently, DNA puffs were observed in other members of the *Sciaridae* family. An increase in the DNA content of the bands at puff sites was shown by Feulgen staining (47), spectrophotometric measurements (48), and by autoradiographic studies of the incorporation of [^{3}H]thymidine (49–51). There are about ten DNA puffs in the salivary gland chromosomes of *R.americana* (*Figure 2*). The genes within the puffs encode abundant proteins that are used by groups of larvae as they spin a large communal cocoon. Early estimates of the degree of gene amplification were somewhat inaccurate, reflecting the technology used. Rudkin and Corlette (48) detected about a 4-fold increase in absorbance of one puffed region in *R.americana* during polytenization in comparison with a 2-fold increase in a region that did not puff. The measurements of Crouse and Keyl (51) on a DNA puff in *Sciara coprophila* showed a 16-fold increase in integrated absorbance compared with a 4-fold increase at a neighboring chromosome site. A more accurate measurement has been obtained utilizing cloned cDNA and genomic probes from two of the major puffs on the C chromosome of *R.americana* (52–54). The cDNA clones were readily isolated in a differential screen of a cDNA library made from salivary gland RNA at a stage at which the puffs are transcriptionally active. Hybridization studies with these probes indicated that the two chromosomal regions undergo 16-fold amplification at the time of DNA puff formation (52–54).

Although DNA puffs are not found in *Drosophila* tissues that contain polytene chromosomes, the phenomenon is probably very similar to the amplification of the chorion genes that occurs in follicle cells and which is discussed in the following section.

3.4.2 Chorion genes

Perhaps the best studied example of developmentally regulated gene amplification concerns the two chorion gene clusters in *D.melanogaster*. The chorion, or egg shell, of *Drosophila* is synthesized around each oocyte by the surrounding follicle cells. The synthesis of the chorion is an extremely rapid process with the mRNA accumulating within about 1 h and almost all of the protein being synthesized within 2 h. In order to achieve this large amount of protein synthesis the clusters of six chorion genes on the X chromosome (*Figure 3*) and the cluster of four on the third chromosome (*Figure 4*) are amplified to a maximum of 15-fold and 60-fold,

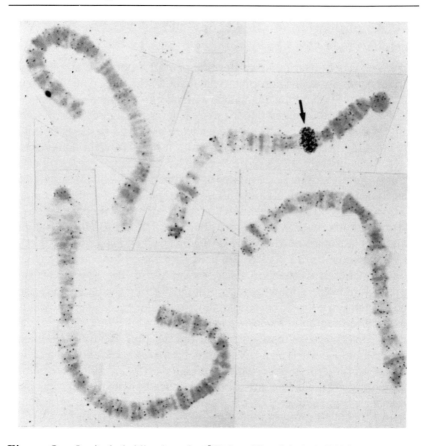

Figure 2.   *In situ* hybridization of a [³H]thymidine-labeled cDNA clone to the DNA puff at the chromosomal region C8 of *Rhynchosciara americana*. The cDNA corresponds to an abundant mRNA from the salivary glands of fourth instar larvae. It is one of several RNAs that encode proteins of the communal cocoon made by groups of larvae. The position of hybridization, marked by the arrow, is to a conspicuously enlarged chromosomal region that has undergone 16-fold amplification by comparison with its flanking sequences. (Reproduced from ref. 52.)

respectively. Mutations that affect chorion protein synthesis include a set that disrupts the process of gene amplification (55,56), resulting in the production of thin egg shells.

The follicle cells are themselves polyploid, ceasing division during oogenesis and attaining an average ploidy of 45C. Amplification of the chorion genes in the follicle cells begins around the time of the last round of endoreduplication of the total genomic DNA. Spradling (57) has analyzed the extent of amplification in the sequences flanking the chorion genes, and showed that a gradient of amplification spreads bidirectionally over a distance of 80 – 100 kb, being maximal in the region of the chorion gene cluster (*Figures 3* and *4*). There is no change in the organization of

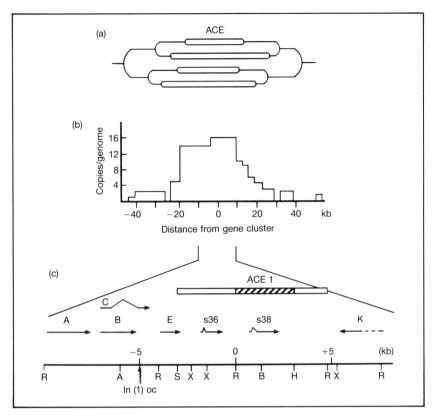

Figure 3. The organization of the amplified chorion gene cluster at the cytogenetic region 7F1,2 on the X chromosome of *D.melanogaster.* (a) The model for chorion gene amplification by the multiple reinitiation of chromosomal replication from a site in the vicinity of the chorion gene cluster. This model is supported by direct observations in the electron microscope (see *Figure 5*), and by measurements of gradients of DNA amplification over 80 kb of the genome as indicated on the histogram (b) (57). The amplification control element (ACE1) is located in the maximally amplified region. (c) This shows the 18 kb central region in greater detail. Restriction sites are as follows: R, *Eco*RI; S, *Sal*I; H, *Hpa*I; A, *Ava*I; X, *Xho*I; B, *Bam*HI. Map distance (kb) are given with respect to an *Eco*RI site indicated at 0. Seven transcription units are shown. A, B, C, E, s36, and s38 are follicle cell-specific RNAs expressed only during the time of chorion formation. Transcript K is expressed in nurse cells but not follicle cells. The position of the breakpoint in the chromosomal inversion *In(1)ocelliless* is indicated by the wiggly arrow. The hatched box indicates sequences that may be essential for ACE1 function. The open boxes indicate sequences that can induce amplification in combination with the hatched region. (Reproduced from ref. 59.)

these genomic sequences as judged by the patterns of restriction endonuclease cleavage. These observations are consistent with a model in which gene amplification results from rounds of localized chromosomal replication with multiple replication forks spreading bidirectionally into the flanking sequences. The model has received support from direct

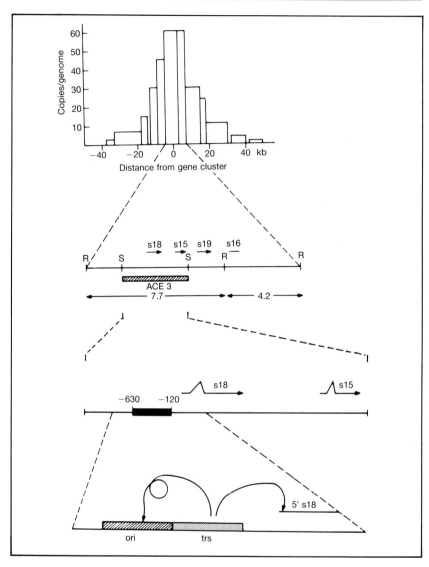

Figure 4.   The organization of the third chromosome cluster of chorion genes of *D.melanogaster*. The upper panel shows the gradient of amplification in about 100 kb of DNA from measurements made on egg chamber DNA at stage 13 of oogenesis when maximum levels of amplification have been reached (57). The coordinates are with respect to the 5′ end of the s18 transcript. The four chorion genes are s18, s15, s19, and s16. The amplification control element, ACE3, was originally mapped to the 3.8 kb *Sal*I fragment indicated by the cross-hatched box. (61). *Sal*I and *Eco*RI cleavage sites are indicated by S and R, respectively. The lower panel shows a more detailed map of the ACE3 element (solid box) on the 3.8 kb *Sal*I fragment (63). This region contains a *cis*-acting transcription regulatory sequence (trs) required for transcriptional control of the s18 gene and also postulated to act in *cis* upon the replication origin (ori). The curly arrows indicate these *cis* interactions (36). (Reproduced from ref. 63.)

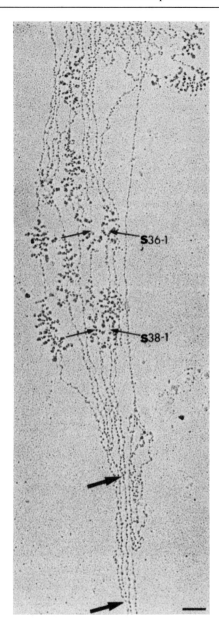

Figure 5.   Electron micrograph of chromatin from follicle cells of *D.melanogaster*. In the lower half of the figure, two replication forks can be seen (large arrows), indicative of the multiple firing of replication origins within this region of the genome. Actively transcribed regions of chromatin (small arrows) are seen in association with these 'over-replicated' regions. Measurements of the spacing of these transcribed regions suggest that they correspond to the chorion protein genes s38 and s36 of the X chromosome cluster. The bar is 0.2 μm. (Reproduced from ref. 58.)

observations in the electron microscope that reveal multiple branched chromatin fibers around transcriptional units that appear to be those of the chorion genes (*Figure 5*). These structures are not seen in spreads of DNA from either early stage egg chambers or embryos (58). The transcribed genes, identified by their size and relative spacing, are seen in close proximity to replication forks upstream and downstream of the transcribed genes. This is a dramatic example of how the controls that prevent reinitiation of DNA replication within a single cycle of chromosomal DNA synthesis in embryonic or other diploid cells (Section 3.2) have to be over-ridden in the chorion gene clusters. It would seem that replicons in other chromosomal regions of follicle cells are unable to respond to replication signals after this stage of development. A considerable body of evidence favors the existence of specialized elements that control chorion gene amplification in follicle cells.

3.4.3 cis-regulation of chorion gene amplification

One of the first examples of a mutation affecting chorion gene amplification was a rearrangement of the X chromosome known as *ocelliless*. This is a chromosomal inversion (see *Figure 3*) in which the DNA proximal to the breakpoint retains its ability to amplify (but at a reduced level), whereas the DNA distal to the breakpoint does not amplify (55). This was the first indication of *cis*-acting sequences in the vicinity of the chorion gene cluster that were required for gene amplification. The development of techniques for the germline transformation of *Drosophila* using vectors based on transposable P-elements has enabled the *cis*-acting sequences to be localized. In such experiments, the transforming DNA becomes integrated into the *Drosophila* genome at random positions. Restriction endonuclease cleavage fragments from either the X or third chromosome clusters have been reintroduced into flies and their amplification in follicle cells assayed. In such a way, the amplification of the X-linked chorion gene cluster has been shown to require a region upstream from the S38 chorion gene (59) (*Figure 3*). This element of DNA, the amplification control element 1 (ACE1) is situated in a 4.7 kb segment of genomic DNA. Although transcription of the S38 gene is not required for amplification, a 467 bp region upstream from the start of the S38 transcription is essential both for correct developmental regulation of transcription and for amplification. Close to this essential region is a repeated motif AATAC, present in eight perfect copies and four imperfect copies. A transposon from which all but three imperfect repeats have been deleted can undergo amplification whereas, when the remaining repeats are deleted, amplification does not occur. Deletion of these repeated elements does not affect transcription of the S38 gene, however, which is under the control of a separate set of sequences (60). Although the *cis*-acting elements required for transcription of the S38 gene can be separated from those needed for amplification of the gene cluster, the precise interrelationships between these control elements remain to be established.

The chorion gene cluster on the third chromosome has been studied in greater detail than the X chromosome cluster since its higher level of amplification makes the experiments technically easier. As indicated above, the approach has been to introduce segments of the gene cluster back into the germ line of flies by P-element mediated transformation. In their initial studies on the amplification of the third chromosome chorion gene cluster, de Cicco and Spradling showed that ACE3 lies within a 3.8 kb DNA fragment that can amplify at ectopic chromosomal sites (61) (*Figure 4*). This contains the two chorion genes S18 and S15 together with the elements necessary for the correct developmental control of their transcription. The amount of amplification is very sensitive to the position of the P-element transposon within the chromosomes; the transposons rarely amplified to normal levels and at some sites no amplification occurred at all. In those cases where there was no amplification, the genes were found to amplify when remobilized by hybrid dysgenesis and inserted at alternative sites. This demonstrates that position effects can lead to the inhibition of amplification. The phenomenon could be due to interactions with elements regulating the local replication of DNA on the new chromosomal sites which may in turn be influenced by the position of particular chromosomal regions within the nucleus. If the P-element transposon carrying the minimal element required for DNA amplification is flanked by sequences from the chorion gene cluster, then the effect of chromosomal position upon amplification is minimized. Unfortunately, it can be difficult to interpret such data and to know whether these naturally flanking sequences are merely acting as buffers to dampen the inhibitory effects of chromosomal DNA, or whether the flanking DNA is stimulating amplification *per se* (61 – 63). It has been suggested by Delidakis and Kafatos (62) that several control elements within the third chromosome gene cluster regulate replication from an origin that is either close to or within ACE3. Thus, deletion of the sequences between S18 and S15 reduced the level of amplification whereas removal of sequences upstream from S16 had less effect. Similarly, Orr-Weaver and Spradling (63) found that a deletion upstream of S15 strongly reduced amplification.

The region essential for the amplification of the third chromosome gene cluster has been further mapped between positions -630 and -120 upstream of the S18 gene (63) (*Figure 4*). These studies also revealed that this region contains an element that is necessary for the transcription of the S18 gene. Germ line transformation has also been used to identify three regions upstream of the S15 gene that regulate its transcription (63). One of these is an activator region containing the sequence TGACGT, a motif found upstream of chorion genes from both *D.melanogaster* and the silk-moth, *Bombyx mori*. The silk-moth sequence is known to be required for the correct tissue-specific expression of *Bombyx* chorion genes introduced into *D.melanogaster* by P-element-mediated transformation (64).

A mutation in the hexamer sequence abolishes all S15 gene expression in *D.melanogaster*.

In a recent study, Spradling and his colleagues have gone on to demonstrate that removing the sequences between -310 and -630 upstream of the S18 gene eliminates amplification and yet does not effect the developmental regulation of transcription in several transformed lines (personal communication). It thus appears that the control elements for transcriptional regulation are distinct from those required for amplification. A detailed analysis of the sequences in the region between -310 and -630 identifies a 60 bp element that is critical in regulating amplification, although there also appear to be functionally duplicated control elements that are able to complement weakly a deletion of this element. It may indeed be the case that there are additional regions outside the ACE3 element that are required to enhance the degree of amplification, as suggested by recent work from Kafatos' group (personal communication).

It is not clear whether the ACE element itself contains a replication origin. Spradling's group have made transformants carrying either nine or 18 tandemly arranged copies of the minimum ACE element. Flies carrying these constructs amplify the DNA at the correct developmental stage but only at a very low frequency, suggesting that the ACE3 element might only be a regulator and not contain the replication origin. This question should receive a definitive answer from studies being carried out by both Spradling's group and Kafatos' group to map replication origins using the two-dimensional gel techniques developed by Brewer and Fangman (65).

3.4.4 Trans-regulation of chorion gene amplification

The laboratories of both Kafatos and Spradling have undertaken a search for genes encoding proteins that act in *trans* to regulate chorion gene expression. This has involved looking for female sterile mutants that affect egg morphology. Several genes have now been identified which map at sites far removed from either chorion gene cluster and which appear to be required for chorion gene amplification (56,66). It seems likely that some of the genes encoding *trans*-acting factors may be specific to chorion gene amplification *per se*, whereas others may represent genes that are essential for more general aspects of chromosome behaviour. The amplification-deficient mutant K451, for example, has been shown to be allelic to *mus101* (B.Baker, personal communication; J.M.Axton and D.M.Glover, unpublished data). *mus101* was first selected on the basis of its increased mutagen sensitivity, and subsequently discovered to have a specific effect upon the condensation of heterochromatic regions of chromosomes in mitotically divided cells (67). The mutation might, therefore, affect proteins required for chromosome condensation *per se* or alternatively, this phenotype could be a secondary consequence of an effect upon DNA replication. It is often difficult to separate cause from effect in such phenotypic observations. Nevertheless, it may well be that the effect of *mus101* upon chorion gene amplification is one of several pleiotropic consequences of a general perturbation of chromosome architecture. The molecular and genetic characterization of these and other *trans*-acting genes over the coming years will undoubtedly increase our understanding of the regulation of chorion gene amplification.

4. Amplification of genes mediating resistance to toxic agents in whole organisms

In unicellular organisms, amplification is a common mechanism for generating variant cells that survive in the presence of a toxic drug or that have the ability to grow well in particular media. For reviews and earlier references concerning amplification in prokaryotes and yeast, see Anderson and Roth (68) and Stark and Wahl (8). Amplification of the gene encoding the bifunctional protein thymidylate synthetase – dihydrofolate reductase has been particularly well studied in antifolate-resistant *Leishmania*. The amplified DNA, present in about 100 copies per cell as a circular dimer, appears to be derived from the chromosomal sequence by replication of a limited region (69). Tunicamycin-resistant *Leishmania* carry amplified copies of the *N*-acetylglucosamine-1-phosphate transferase gene responsible for resistance also on a small extrachromosomal DNA molecule (70). This suggests that the mechanism may be quite general for *Leishmania* and perhaps for other protozoa.

Formation of drug-resistant single-celled organisms by means of amplification is perhaps not too surprising in the light of experience with mammalian single cells in culture (Section 5). However, these results do not really prepare us to appreciate that entire multicelled organisms such as insects can also achieve drug resistance by means of gene amplification. A strong early clue was provided in 1979, when Devonshire and Sawicki (71) showed that a series of insecticide-resistant aphids, from independent field isolates, had increased levels of an insecticide-hydrolyzing esterase and that the levels of enzyme formed a geometric series, suggesting successive rounds of gene duplication. In 1986, Mouchès *et al.* (72) showed that amplification of an esterase gene gave insecticide resistance in mosquitoes, with increases in gene copy number of at least 250-fold. More recently, Field *et al.* (73) have shown conclusively that esterase gene amplification is in fact responsible for insecticide resistance in aphids and have also obtained preliminary evidence concerning the molecular events. It is very interesting that loss of the amplified arrays in aphids can be catastrophic (74). Further study of these systems should shed light on the mechanisms involved, which may well be relevant to insecticide resistance in other species. Remarkably, Prody *et al.* (75) have found a human family in which a defective gene encoding serum butyrylcholinesterase has been amplified about 100-fold, at or near the normal site for the single-copy gene on chromosome 3. It seems likely that selection took place early in oogenesis, where the defective gene rendered cells highly susceptible to poisoning by agricultural organophosphorus insecticides.

5. Amplification in mammalian cells in culture

5.1 Amplification in drug-resistant cell lines

Most of our information concerning mechanisms of amplification of mammalian DNA comes from studies of drug resistance in cultured cells. Since the last tabulation of agents known to select for such amplification (9), most of the cases listed as 'likely' or 'possible' have now been discovered to be due to amplification. In one case, only the α-subunit of the oligomeric enzyme Na/K-ATPase is amplified in ouabain-resistant cells (76). In another, cells resistant to increasing concentrations of hydroxyurea have been shown to undergo a complex set of changes in which there is first over-expression without amplification of the M2 subunit of the oligomeric enzyme ribonucleoside diphosphate reductase. This is followed, first by amplification of the M2 gene, then by other events which do not involve amplification, and finally by amplification of the M1 gene in the most highly resistant cells (reviewed in ref 77). Amplifications of the genes for thymidylate synthetase (78), inosine monophosphate dehydrogenase (79) and AMP deaminase (80) have now been proved to mediate resistance to the appropriate drugs. In an interesting recent case, amplification of genes encoding the α class of glutathione S-transferases gives resistance to nitrogen mustards in Chinese hamster ovary (CHO) cells (81). This is the first example in which resistance to an alkylating agent is known to be mediated by amplification.

5.2 Multidrug resistance

In multidrug resistance (MDR), cells selected with only one of a set of seemingly unrelated drugs (e.g. doxorubicin, vinca alkaloids, dactino-mycin, colchicine, or taxol) are resistant to all or most members of the set. The basis, now well understood, is over-expression of a gene (*mdr*-1) encoding a protein (p170 glycoprotein) that functions as an ATP-dependent efflux pump of broad specificity in the plasma membrane of resistant cells (82). MDR has been the object of much activity and has been reviewed several times in recent years (83–87). Initially, amplification was found to be responsible for MDR in tissue culture cells (88–90) but more recently it has been appreciated that low level resistance is often due to elevation of *mdr*-1 mRNA without amplification (91,92). In retrospect this is not surprising, since the *mdr*-1 gene is normally expressed at high levels in a few tissues (liver, kidney, and intestine), presumably because it normally has a role in detoxification *in vivo*. Thus, mechanisms that suppress accumulation of *mdr*-1 mRNA in tissues where it is not normally expressed can be lost, followed by amplification to give the higher levels of mRNA needed to resist high drug concentrations. Although MDR is a major cause of clinical drug resistance, it is not yet clear whether amplification has a substantial role in mediating such resistance *in vivo*.

Several other genes are typically co-amplified with *mdr*-1 and consequently over-expressed, but it has become clear that over-expression of *mdr*-1 alone is responsible for resistance (93,94). Interestingly, Choi *et al.* (95) have shown that the different ranges of cross-resistance often observed in MDR cells selected with different agents are likely to be due to spontaneous mutations in a single MDR gene that mediates resistance (*mdr*-1) rather than due to over-expression of a different member of an *mdr* gene family with a different specificity. In at least one case, it has been possible to find a variant of doxorubicin (one of the MDR family of drugs) that is equally effective on drug-sensitive and MDR cells (96). Hopefully, many additional ways to circumvent MDR *in vivo* will be found.

5.3 Parameters of amplification

5.3.1 Steps of selection and primary events

Amplifications are often selected in distinct, operationally defined steps of increasing drug concentration. A primary event takes place in a single cell to generate extra copies of the target gene. Secondary events, not necessarily selected for by the drug, may then follow during growth of the initial cell to a clonal population sufficiently large to be analyzed. Schimke (11) has pointed out that resistant clones often grow slowly and the growth rates often increase only after prolonged selection. One suspects that more than one process occurs before the first-step clonal population is established.

5.3.2 Rates and frequencies

Although it is possible to stimulate amplification events (see Section 7), they do occur spontaneously, quite independently of drug treatment (10, 97–99). The frequency of amplification, in the range 10^{-4} to 10^{-6} for most cell lines, is determined by three parameters: the number of new amplifications per cell per generation; the rate at which amplified DNA is lost from cells; and the relative growth rates of cells with and without amplifications. In contrast to frequency, the rate of amplification is a simple parameter, although more difficult to determine experimentally. As expected, both rate and frequency are functions of drug concentration, since low drug concentrations will kill cells with a low degree of amplification while allowing those with more copies to survive. It is less intuitively obvious, but true, that frequencies and rates may vary quite differently with drug concentration (8,99).

5.3.3 How to study new and independent events

Even though not usually done, it is a relatively simple matter to obtain clonal populations of resistant cells which do not include pre-existing amplifications. This is especially important when the structure of the amplified DNA or the properties of the resistant populations are to be related to amplification mechanisms, since one does not want to study cells or structures which may have changed in unknown ways during prolonged growth in culture. To obtain new and independent amplifications, separate populations, small enough so that few contain a pre-existing mutant (e.g. 10^3 cells per population if the frequency is 10^{-5}), are plated out. The populations are then grown *without* selection to increase the number of cells by 100- to 1000-fold and then placed in a drug. If a population gives rise to only one or a few resistant colonies, the amplification event must have occurred only one or a few cell divisions before selection was imposed.

5.4 Amplification as an abnormal event

It is widely believed that most normal mammalian cells rarely amplify their DNA, based largely on the failure to observe cytological manifestations of amplification [extended chromosomal regions (ECRs) and double minute chromosomes (DMs)] in normal cells, in contrast with their frequent observation in cell lines and tumors (1). It follows that high frequency amplification (usually in the range 10^{-4} to 10^{-6} at a single locus) has somehow been stimulated in cell lines and other abnormal cells. Increased rates of amplification have been correlated with tumorigenicity when cell lines with different tumor-forming potential have been compared (100,101). However, in another case, this correlation was not found (102). Drug-resistant variants of normal cell strains have usually been very difficult to select. For example, it has not been possible to select methotrexate (MTX)-resistant cells from a large number of normal human mammary epithelial explant cells in parallel with ready selection of such cells from a mammary carcinoma cell line (H.Smith and G.R.Stark, unpublished data), nor to select *N*-(phosphonacetyl)-L-aspartate (PALA)-resistant cells from a large number of normal human keratinocytes in culture (F.Watt and G.R.Stark, unpublished data). However, there is one report of c-Ha-*ras* amplification in normal human diploid fibroblasts *in vitro* (103) and of amplifications of HPRT (104) and HLA (105) genes in normal human lymphocytes. Clearly, there is a need for more of these difficult experiments to be attempted.

6. Mammalian gene amplification *in vivo*

6.1 Amplification of oncogenes

There is as yet no example of developmentally programmed amplification in mammals. Thus, mammalian amplification is known only as an abnormal event in isolated cells or in tumors. Although there are only a few instances of amplification in drug-resistant tumors (reviewed in ref. 9), there are many cases in which cellular oncogenes are found to be amplified in advanced tumors (106 – 108), leading to the logical conclusion that their over-expression contributes in a major way to tumor progression. Acquiring the abnormal ability to amplify many different subregions of the genome at high frequency can greatly benefit an emerging tumor cell population, by providing a means to generate many different types of variant cells, some of which can evade the immune response, become metastatic, and so-on. Also, it seems likely that amplification is but one manifestation of more fundamental defects that lead to other chromosomal abnormalities such as deletion, inversion, translocation, and possibly even chromosome loss (see Section 7.4). These events also increase the probability of diversity, with advantage to the tumor cell population.

In some types of tumors, particular oncogenes are found to be amplified in a substantial percentage of the cases examined, often as high as 20 – 50% (see reviews 9 and 106 – 108 for references before 1987). The main correlations are N-*myc* in neuroblastomas and retinoblastomas; c-*myc* in breast cancers (109,110); the entire *myc* family in lung cancers (111,112); c-*erb*B-2 (also known as *neu*) in breast cancer (110, 113 – 116) and tubular adenocarcinomas of the stomach (117); c-*erb*B-1 (epithelial growth factor receptor) in gliomas (118 – 121) and breast cancer (122); c-Ki-*ras*-2 in embryonal carcinomas (123); *int*-2 [and probably also *hst*, since they are closely linked (124)] in breast cancers and squamous carcinomas (125).

Amplification of a particular gene occurs rarely even in abnormal cells and would not be present in all cells of a population unless selected. Although the degree of correlation of amplification of a particular oncogene with a particular tumor type is already striking, we must consider that the frequencies determined so far are underestimates, for two reasons:

(i)   In certain tumors, any of several oncogenes can be amplified; especially breast cancer, where c-*myc*, c-*erb*B-2, c-*erb*B-1, and *int*-2 (or *hst*) are each found to be amplified in a significant fraction of the cases examined (see refs above). Therefore, the great majority of breast cancers will carry an amplification of at least one of the above set. Amplification of alternative oncogenes is seen in other cases as well. Furthermore, since there must be oncogenes that are as yet undiscovered, we are probably underestimating the number of such situations.

(ii) A low degree of amplification, which may carry a significant selective advantage, can easily be missed unless the assays are carried out with very careful internal controls. Furthermore, contamination of tumor samples with non-tumor cells makes estimates of the degree of amplification low and probably causes some examples to be missed.

6.2 Correlation of amplification with metastasis

The degree of amplification of a cellular oncogene often correlates with the aggressiveness of metastatic behavior of the tumor [for references, see the reviews cited above and also two recent examples of c-*myc* (126, 127)]. The idea that genetic instability is important in tumor progression is, of course, not new, and has been discussed critically in several excellent recent papers (128–133). With respect to amplification, it is interesting that in one case (134) there was a close correlation between the increased ability of a cell line to generate drug-resistant variants and more metastatic variants. However, such a correlation was not observed when four cell lines with increased rates of amplification (Section 7.2) were tested for increased rates of metastasis (135). Increased rates of amplification have been correlated with tumorigenicity when cell lines with different tumor-forming potential have been compared (100,101). Also, increases in the two major cytogenetic manifestations of amplification, ECRs and DMs, have been correlated with increased ability to form tumors (136), to metastasize (137), or to invade (138). The nature of the genes amplified was not analyzed in any of these cases.

6.3 General methods to detect or clone amplified genes

When the genes to be assayed are known and cloned probes are available, the usual method of analysis has been to hybridize the probes to Southern transfers prepared from isolated DNA. Despite its great utility, this method is not ideal for quantitative analysis of a large number of cell or tissue samples. An alternative slot-blot assay developed recently is extremely sensitive and accurate and entirely does away with the need to prepare DNA samples (139). Also of potential use is a method in which gene copy number can be assayed in many independent colonies of mammalian cells (140).

When cloned probes are not available, amplified DNA can be cloned based solely on its increased relative abundance. The earlier literature on the methods used was reviewed by Stark (9). A major problem is lack of sensitivity; it is still very difficult to clone sequences that are amplified by less than about 20-fold. Nevertheless, there have been some notable recent advances. The in-gel renaturation method developed by Roninson takes advantage of the fact that single-stranded DNA sequences anneal faster when present at higher concentrations and thus are protected from nuclease digestion. The method has been developed further (141,142) and used to clone amplified sequences from adriamycin-resistant human breast cancer cells (143) and human tumor cell lines (144). The gene *gli*, amplified and highly expressed in a human glioma, was cloned by screening genomic clones, originally obtained by the in-gel renaturation technique, with RNA-based probes, to identify the subset of amplified sequences that were expressed (145). The genes *mdm*-1 and *mdm*-2 were isolated from a transformed mouse line by differential screening of a cDNA library for over-expressed sequences, followed by assay of the clones obtained for amplification (146). The new technique of phenol emulsion competitive DNA reassociation has also been used to clone amplified DNA from a human neuroblastoma cell line (147). Continued application of these methods should yield even more new examples of genes amplified in tumors.

7. Stimulation of amplification

7.1 Stimulation by exogenous agents

7.1.1 Virus infection

Amplification of host cell sequences can be triggered by viruses such as polyoma or SV40 when viral origins of replication integrated in host DNA are stimulated by the T-antigens (148). Infection of cells with Epstein – Barr virus (EBV) *in vitro* leads to amplification and rearrangement of the *myc* oncogene (149), a very interesting result in view of the fact that over-expression and rearrangement of *myc* is often found in EBV-positive Burkitt's lymphomas. Amplification units in melanoma cells from two patients show homology with a human papilloma virus and with EBV (150), and amplification of integrated hepatitis B viral DNA, associated with chromosomal translocations in hepatocellular carcinomas (151), has been shown, in one case (152), to include co-amplification of the oncogene *hst*-1.

There are at least two ways in which infection with viruses strongly associated with carcinogenesis, such as EBV, papillomaviruses, or hepatitis virus B, can lead to DNA amplification.
(i) The viruses may cause specific DNA rearrangements which lead to amplification-prone structures.
(ii) There may be selection in tumor cells for over-expression of viral gene products, with coincidental amplification of cellular sequences which flank the integration sites.

7.1.2 Stimulation by chemicals or radiation

Amplification can be induced by a wide variety of agents or treatments which interfere with DNA synthesis or damage DNA, for example, hydroxyurea (153), aphidicolin (154), carcinogens (155,156,157), hypoxia (158), UV radiation (155), and ionizing radiation (159). In rat cells treated with bromodeoxyuridine (BrdU), Pasion *et al.* (160) identified a 4 kb sequence near the prolactin gene that mediates enhanced amplification of this gene or of constructs containing the sequence. It seems very likely that stresses such as those listed above lead to induction of gene expression (156) and that the induced gene products act in a dominant fashion to stimulate amplification (159). Treatment of cells with chemicals or radiation, or infection with certain viruses can also induce the replication of integrated viruses, including flanking cellular DNA. (For two recent examples and references to prior work, see refs 161 and 162.) The acentric chromosome fragments present in irradiated cells may also stimulate amplification (163).

It is not yet clear how stimulation comes about. The early idea that many cells respond to relief from inhibition by reinitiating DNA synthesis within a single cell-cycle (164) has not been supported by later experiments (165 – 168). However, the conclusion that cells do not re-start synthesis within a single S-phase is limited by the sensitivity of the methods employed, and it remains possible that a small fraction of the treated cells do restart in S-phase and can contribute to amplification in response to stress. More recent observations employing flow cytometry of cells pulsed with BrdU (154) indicate that, if DNA synthesis of cells in S-phase is inhibited, their entry into mitosis is delayed after resumption of synthesis, and they contain more DNA than normal G2/M cells. It is now proposed that the relationship between S and M may be perturbed in the stressed cells so that mitosis occurs variably in a second S-phase (11, 169). Furthermore, Sherwood *et al.* (170) have shown that protein synthesis is necessary for the amplification induced by treatment with inhibitors of DNA synthesis.

7.2 Amplificator cell lines

Recently, mutant cell lines with higher rates of amplification (amplificator cells) were selected from Syrian hamster BHK cells. Simultaneous exposure to the two selective agents MTX and PALA gave rise to doubly resistant cells 20 – 200 times more frequently than predicted for the corresponding single drug selections, and the rates of amplification of three additional genes were increased by up to 25-fold in the doubly resistant cells (171). Simultaneous selection of mouse B16 melanoma cells with MTX and PALA also selects doubly resistant cells with an amplificator phenotype (172) and simultaneous exposure of Chinese hamster cells to MTX and adriamycin gave double resistance 10 – 100 times more frequently than predicted (173). Fusion of amplificator BHK cells with monkey cells leads to greatly enhanced amplification of the monkey CAD gene, showing that the amplificator phenotype is dominant (174). The opportunity to work with reasonably stable cell lines altered in their amplification rates opens the possibility to isolate genes responsible for this phenotype. It is not unlikely that synthesis of similar or identical stimulatory proteins can be induced transiently by stress, or more stably by the type of genetic change present in amplificator cells. Preliminary investigation indicates that amplificator cells are hypersensitive to UV light and mitomycin C (E.Giulotto, unpublished data), raising the possibility that a single underlying defect can cause several different phenotypic changes in these cells.

7.3 Amplification and other chromosomal abnormalities

The idea that a common defect underlies amplification and abnormalities such as translocation, inversion and deletion has been put forward in several reviews (1,5,8,12) and especially by Schimke *et al.* (175) with respect to the onionskin model (Section 10.1.1) and by Carroll *et al.* (176) with regard to episome models (Section 10.2.1). Recently Ottagio *et al.* (177) have shown that Chinese hamster cells selected for PALA resistance, which is always mediated by amplification of the CAD gene, have a much higher incidence of chromosome aberrations than do the parental cells. The potential connection between deletion and amplification is especially intriguing. Deletion appears to be a common mechanism for generating thymidine kinase (TK) mutations in human cells (178) and adenine phosphoribosyl transferase (APRT) mutations in CHO cells (179–181), where it is interesting that coordinate amplification–deletion events have been observed (182). Deletions have also been coupled to amplification in normal human lymphocytes at both the HPRT (104) and HLA (105) loci and, as mentioned above, in the amplification of the dihydrofolate reductase (DHFR) gene in CHO cells (B.E.Windle, B.Draper, and G.M.Wahl, in preparation). Intrachromosomal recombinations between highly repeated sequences are a frequent source of low-density lipoprotein (LDL) receptor mutations (183,194). Interstitial deletion is a common mechanism for the loss of tumor suppressor genes associated with malignant transformation (185–189). These observations encourage the speculation that similar mechanisms could generate deletions leading to allele loss and decreased function on the one hand, or gene amplification and over-expression on the other, with the outcome being selected by the particular drugs used.

8. Forms and structures of amplified DNA

8.1 Cytological manifestations: interrelationships of ECRs and DMs

Amplified DNA can often be observed by light microscopy either as an ECR or as extrachromosomal elements, DMs (*Figure 6*). Although the same gene can sometimes be amplified either chromosomally or extrachromosomally in a single population of cells, the two forms do not usually co-exist within the same cell (1,191–193).

Whether DMs and ECRs represent different outcomes of the same type of primary event or result from fundamentally different processes is unclear at present. DMs may form during a primary event and then be replaced by ECRs (e.g. see ref. 194) or, conversely, ECRs may form first and then give rise to DMs secondarily. At first sight, the latter possibility seems to contradict the usual experience that ECRs are stable, based on studies carried out after several steps of selection (8,97) and supported

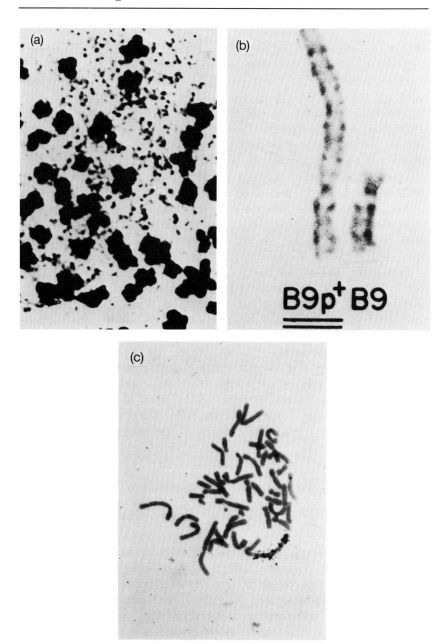

Figure 6. Double minute chromosomes (DMs) and extended chromosome regions (ECRs). (a) A partial metaphase chromosome spread from the human nueroblastoma cell line TR14 (reproduced from ref. 1). (b) Syrian hamster chromosomes B9p⁺ and B9, which carry about 100 copies of the CAD gene following selection with PALA (190). (c) *In situ* hybridization with a CAD cDNA probe of a metaphase spread from the same cells (190).

by the experiments of Cherif *et al.* (193) with a human breast carcinoma line, where the ECR initially present does not break down during growth *in vivo*. However, recent work from several laboratories shows that the newly formed ECRs present after early steps of amplification are lost after a short period of cell growth without selection (see Section 9.2). Evolution of amplified chromosomal arrays, giving rise to new arrays with more copies of a selected gene, may involve either new intrachromosomal or new extrachromosomal primary events. More definitive work is needed, using new techniques to study cells as close as possible to the primary events of different steps, to investigate these different mechanistic possibilities in cells of several species and involving several different genes.

The general impression that it is difficult to observe DMs in Chinese or Syrian hamster cells (99,195) could be due to more frequent reintegration of extrachromosomal forms in these cells compared to mouse, rat, or human cells where DMs are often observed. In two cases, in CHO cells, frequent reintegration has been observed (176,196). It has not been determined in any system whether DMs integrate at the locus of the native gene by homologous recombination, as first suggested by Biedler (197), or whether they integrate more or less at random. Localization of amplified sequences to multiple chromosomal sites in cell lines has been interpreted as being due to translocation of amplified sequences from ECRs (11,198), but it is also possible that integration of extrachromosomal sequences could generate such patterns (176). Reinsertion at the native locus may be quite likely since it is now known that the frequency of homologous integration increases as a strong function of increasing homology between donated and target sequences (199) and, for example, the homologous regions would span more than 100 kb for episomes described below.

The fact that DMs are the most common form of amplified DNA in diverse human tumor samples analyzed immediately after biopsy (S.Benner, G.M Wahl, and D.D.von Hoff, in preparation) could be due to the well documented selection of cells with DMs *in vivo* (191 – 193). Unfortunately, it is not possible to investigate the state of amplified DNA in tumor cells close to primary events, so the mechanistic relevance of these observations is unclear, apart from the obvious conclusion that DMs must be an important intermediate in the formation or evolution of amplified DNA in tumors. When tumor cells with DMs are cultivated *in vitro*, they are soon replaced by cells with ECRs containing the same amplified target gene. However, tumors re-established from such cells again predominantly contain DMs (1,193,200,201). The cell lines grown *in vitro* probably contain a few cells with DMs, and these are then highly selected *in vivo*. As noted above, a *cloned* tumor cell line with an ECR does not give rise to DM-containing tumors (193).

8.2 Characterization of DMs

DMs are paired, circular euchromatic elements with an ultrastructure similar to that of metaphase chromosomes (1,202,203). The smallest DMs visible under the light microscope are probably larger than 1000 kb, and DMs of average size are likely to contain more than 5000 kb of DNA, assuming typical compaction ratios (202). However, molecular analyses lead to estimates of less than 300 kb for the genetic complexity of some DMs (145,204) and, in combination with the common observation that DMs within a single cell often vary dramatically in size [e.g. see Trent *et al.* (201)], suggest that large DMs may originate from small precursors. Evidence obtained in human, mouse, and hamster cells involving amplification of cellular oncogenes and genes mediating drug resistance supports the proposal that DMs do originate from submicroscopic precursors in mammalian cells (reviewed in ref. 13). In one example, the submicroscopic elements were shown to increase in size over time and to form structures microscopically indistinguishable from DMs (176). Interestingly, time-dependent increases in size have also been observed for extrachromosomal circular elements containing amplified thymidylate synthetase-DHFR genes in MTX-resistant *Leishmania* (205). These observations suggest that larger circular elements predominate eventually, perhaps because they segregate with higher efficiency than smaller ones at mitosis, as do synthetic linear yeast chromosomes (206). The DM precursors, termed episomes, are circular and replicate once per cell-cycle (reviewed in ref. 13). The diversity of systems and selective conditions in which episomes have been found indicate that they are often intermediates in gene amplification. The similarity in size between episomes and replicons, and the fact that episomes replicate according to the cellular program, suggest that units of replication may correspond to units of DNA amplification in some cases.

8.3 Characterizaton of ECRs

The DNA amplified together with a selected gene includes not only sequences immediately flanking the gene in wild-type DNA but also more remote sequences, which may or may not be co-amplified in virtually every independent event (80,207–213). During amplification, DNA sequences from different parts of the genome are joined together to form novel structures. The region involved may be very large indeed, and estimates up to 10 000 kb have been made for the amount of DNA co-amplified with each copy of the target gene in the first step (213,214). In many highly resistant cell lines, selected in several steps, most of the amplified DNA is represented by a single rearranged structure formed relatively early in the amplification process (145,215–221). The chromosomal location of amplified DNA has usually been investigated after several steps of amplification. In a recent example, the amplified DNA in MDR Chinese hamster cells was studied and it was concluded that the process was more complex than retention of the amplified array in the position of the original single-copy gene (222). These authors also discuss previous work concerning chromosomal rearrangements involving amplified DNA.

8.4 Novel joints in amplified DNA

Early expectations about the structure of chromosomally amplified sequences have been overturned by recent studies in which junctions between amplified domains were characterized. These novel joints give rise to restriction patterns not found with unamplified DNA. Based on the structure of amplified DNA in bacteria and yeasts, it was expected that chromosomally amplified mammalian DNA would be organized as head-to-tail repeats. In contrast, as first described by Ford *et al.* (223), many amplified novel joints have a remarkable structure (*Figure 7*); two palindromic arms, which may extend for many hundreds of kilobases, flank a much smaller non-palindromic center of variable size (200 – 1000 bases). The cross-over site is asymmetric, at the junction between one of the palindromic arms and the unduplicated central sequence (182,196, 216 – 218,224,225). Amplified DNA often contains very large inverted repeats, as shown by snap-back experiments (226) and by chromosome walking (217,224). In one cell line, 80% of the highly amplified DHFR DNA was present as many copies of a 500 kb palindrome (217).

Amplified novel joints, a common feature of highly amplified DNA (80, 207 – 212), can arise by secondary amplification of a single-copy joint formed in an earlier step. It can be clear that this has occurred when a series of cloned cell lines are analyzed (80,214,215,227). However, amplified novel joints also arise *de novo* in a single step (227,228) and in two cases they are inverted and present in every extra copy formed (196,224). Simultaneous formation and amplification of inverted joints was also observed in a case where there was no selection for amplification (182). In different mutants derived from the same parental line and selected for amplification of the same locus, both coupled and sequential formation and amplification of novel joints have been observed (215,224).

Although it was assumed initially that independent recombinations (or strand-switches: see below) would occur randomly in amplified DNA (207,

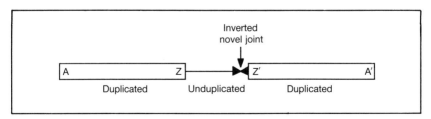

Figure 7.   Schematic structure for the region surrounding a typical inverted novel joint associated with an amplification event. A to Z and A′ to Z′ represent the 5′ to 3′ sequence of a single strand of DNA and of its complement, respectively.

229), an unrearranged DNA sequence which behaves as a recombinational hotspot has been found near the AMP deaminase locus and analyzed (228). The sequence was highly enriched in Alu-like repetitive sequences, in palindromes, and in long AT-rich segments; features known to favor recombination (literature cited in ref. 228; reviewed in ref. 12). Two novel joints formed at this hotspot were sequenced. One is inverted and the other has not yet been characterized fully. In both cases, the breakpoints lie at the tops of stable stem-loop structures. Eight junctions and their unrearranged counterparts from other systems have also been sequenced and five are inverted. In two (230,231), long AT-rich regions surround the breakpoint; in the third (225) the breakpoint lies in a palindromic sequence; in two others (182; J.C.Ruiz and G.M.Wahl, in preparation) no sequence peculiarity was observed. Although the other three joints are not yet fully characterized, the cross-over sites are AT-rich (231). Taken together, these results indicate that the events are generally not homologous and do not reveal an absolute requirement for a particular type of sequence at which an event is likely to occur, although AT-rich elements are often involved.

9. Evolution of amplified DNA

9.1 Preferential reamplification

If the amplified DNA in a first-step clonal population is extrachromosomal, increasing the drug concentration in a second step will simply select for cells with more of the same element, so there is no need to invoke a new primary event. However, if the amplified DNA is intrachromosomal in the first-step population, a second and distinct primary event, taking place in a single cell, is required to generate the additional copies needed for survival at a higher drug concentration. Although, in principle, such a second event could occur at an unamplified allele of the selected gene, it usually seems to involve the already amplified array present after the first step. Amplified DNA contains structures not present in unamplified DNA, such as novel joints and inverted repeats. Features of these unusual structures may allow the primary event of a second step to proceed by a mechanism not available for the primary event of a first step.

In the AMP deaminase system, a single copy of a complex rearrangement was present in a first-step mutant and only the unit identified by this rearrangement was reamplified in five out of five independently derived second-step mutants. Thus, the choice of sequence to be reamplified was anything but random in this case. Results such as these are explained best if rearrangement of DNA creates or bring close to the selected gene sequences which act in *cis* to favor amplification (80 and M.Debatisse, unpublished data). Preferential amplification has been observed for transfected CAD genes, where there is clearly a position effect (232). In several cases, preferential amplification of transfected sequences involved formation of episomes from the original chromosomal structure at high frequency (197,234).

9.2 Stability of amplified DNA

In drug-resistant cells, rapid loss of gene copies in the absence of selection is common when the amplified DNA is extrachromosomal (e.g. see ref 194). In contrast, relatively stable, highly amplified chromosomal arrays have been noted in several systems (97,234,235) and, in some cases, little loss was observed even after culture for more than a year without drugs. Therefore, it was surprising to find recently that chromosomally amplified genes can be lost rapidly from first- and second-step mutants in the absence of selection (10,227). Such loss could occur either by deletion within an amplified array or by loss of an entire chromosome. It has also been shown that several different specific regions of DNA usually co-amplified with a selected gene in first-step populations were often lost after second or third steps, together with preferential reamplification of a smaller part of the initial array (213,215,227). Very highly amplified chromosomal arrays of CAD (227) or DHFR (218) retain only a small amount of co-amplified DNA flanking the selected genes, indicating that loss of initially co-amplified DNA from ECRs is progressive and extensive.

10. Mechanisms of amplification in mammalian cells

Studies of developmentally controlled gene amplification reveal that different mechanisms can generate amplified DNA (Sections 2 and 3). The timely over-production of gene products is satisfied through operation of a different mechanism in each of several well documented systems, a likely manifestation of 'molecular tinkering' (236), utilized by evolution to solve basically similar problems. It seems reasonable to consider that cultured mammalian cells and tumor cells *in vivo* have achieved the ability to use several different amplification mechanisms. Such a possibility is consistent with the varied structures and localizations of amplified sequences in the systems under study. It is also possible that a single basic mechanism might lead to different outcomes, depending only on the probabilities of how structures generated in a primary event are resolved. For example, recombination in the onionskin structure proposed by Schimke and co-workers (175) or within a replication intermediate (13) might give rise to an ECR directly or to episomes (175). The episomes might then reintegrate or remain extrachromosomal, perhaps evolving to form DMs. It is also possible for mechanistically different primary events to occur at different loci within the same cells (depending on *cis* elements), at the same loci in different cells (depending on *trans*-acting factors) or during different steps of amplification (depending on unusual structures generated in earlier steps). As for developmentally controlled amplifications, a single mechanism could predominate eventually, depending on the nature of rearranged structures created by chance during early steps of amplification.

Mechanisms can be subdivided into two major classes. In the first, over-replication within a single cell-cycle is responsible for generating the extra copies. Two subclasses can be distinguished, depending on whether the over-replication is due to multiple initiations or is caused by particular recombination or strand-switching events. In the second major class, unequal segregation is involved and the control of replication within each cell-cycle is normal. In these cases, copy number can increase no faster than the power function 2, 4, 8, 16 . . . in successive cell divisions.

10.1 Replication-driven mechanisms

10.1.1 Multiple initiation in a single cell-cycle

The onionskin model (*Figure 8*), suggested by Tartof (237) and Bullock and Botchan (238), has been verified experimentally in the case of the developmentally controlled amplification of chorion genes in *Drosophila* follicular cells (Section 3.4). In mammalian cells, the gradient of copy number predicted by the model has been observed for amplification of transfected DNA (239). Generating an onionskin requires multiple reinitiation of DNA synthesis within a single cell-cycle. However, early evidence in favor of this event (163) has been challenged by later work (Section 7.1.2). An onionskin structure can be resolved into head-to-head or head-to-tail tandem arrays of amplified units by independent recombination events which result in single copy joints. The onionskin model can even account for the formation of amplified inverted joints during a single amplification event if the unit is formed early and then amplified each time replication is reinitiated in the onionskin. As pointed out by Schimke *et al.* (175), the onionskin model is flexible enough to account for almost any type of molecular product, including extrachromosomal DNA.

10.1.2 The extrachromosomal double rolling circle model

This model was proposed initially to explain amplification of the 2 μ yeast plasmid. During replication of this circular plasmid, which contains a single replication origin, amplification is achieved when a cross-over between one of the newly replicated sequences and a yet unduplicated sequence inverts one of the replication forks. The two forks, which normally converge, now follow each other indefinitely and a multimeric array of the original plasmid is produced (240,241). Passananti *et al.* (225) proposed a model based on the general features of this process: after a single round of local DNA replication, a circle containing an inverted duplication is

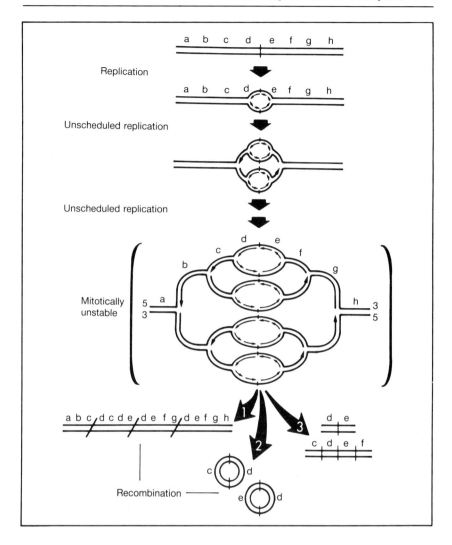

Figure 8.   The onionskin model for amplification involving multiple initiations of replication in a single cell-cycle (reproduced from ref. 8). Unscheduled DNA synthesis and recombination can generate linear intrachromosomal amplified arrays, extrachromosomal circles, or extrachromosomal linear duplexes. Bidirectional replication at an origin generates a bubble that can undergo further rounds of unscheduled DNA replication, resulting in a nested set of partially replicated duplexes. Note that there are only two contiguous chromosomal strands. It is possible for linear duplex DNA to become detached from the structure if two replication forks can approach one another very closely (pathway 3). Recombination within the same duplex could generate extrachromosomal circles (pathway 2), while multiple recombinations among different duplexes could resolve the structure into an intrachromosomal linear array (pathway 1). The example shown depicts a linear array with sequences in the middle amplified more than those at the ends. The ratios are: 1a:1b:2c:4d:3e:2f:1h.

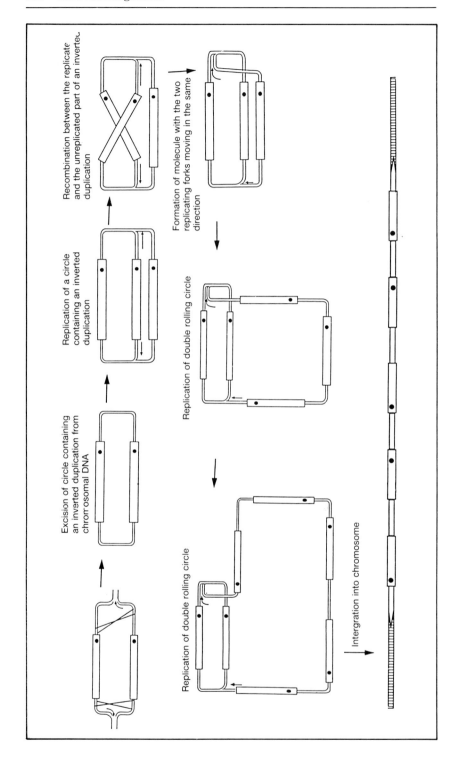

Recombination between the replicate and the unreplicated part of an inverted duplication

Replication of a circle containing an inverted duplication

Excision of circle containing an inverted duplication from chromosomal DNA

Formation of molecule with the two replicating forks moving in the same direction

Replication of double rolling circle

Replication of double rolling circle

Intergration into chromosome

excised from the chromosome by two recombinations and then amplified as a double rolling circle (*Figure 9*). In this case, the lack of a site-specific recombinase might be overcome because the very large size of the inverted repeats might favor events mediated by general recombination. As shown in *Figure 9*, an extrachromosomal circle containing two copies of the same replication origin is formed. Only one can be used during the amplification cycle since there is no simple way to form a double rolling circle with four replication forks. Alternatively, the two initial non-homologous recombination events might exclude one of these two origins from the circle.

10.1.3 The chromosomal spiral model

This new model was developed to explain the novel sequence properties of the recombination hotspot analyzed in the AMP deaminase system. It postulates neither replicon misfiring nor a chromosome break (224). Rather, it considers that, as suggested previously by Nalbantoglu and Meuth (182), an inverted duplication can be generated if replication switches strands and proceeds around the fork as it progresses through a region enriched in palindromic sequences (*Figure 10*). If the unduplicated stem-loop is excised, the resulting structure, which mimics a bona fide replication fork, can prime repair replication, now proceeding toward the already duplicated DNA. The outcome of two such excision – repair processes on both sides of the origin in a replication bubble will be that two replication forks follow each other, yielding a multimeric array of inverted duplications. This model can account particularly well for cases where the amplified array is located at the original chromosomal locus of the selected gene and is organized as a monotonous tandem array of inverted units.

10.2 Segregation-driven mechanisms

10.2.1 The deletion plus episome model

All episomes studied thus far contain functional replication origins. Two different mechanisms can generate circular molecules containing origins: re-replication (175) and recombination across a replication loop (176,225). Recombination across a replication loop leads to a chromosomal deletion

Figure 9.   A model for amplification involving inverted duplications, based on the Futcher model (240) for amplification of the 2 μ circle of yeast. The drawing is adapted from Passananti *et al.* (225). A circular molecule containing an inverted duplication (rectangle) harboring the gene to be amplified (dot) is excised from the chromosome, yielding also a chromosomal deletion (not shown). After the inverted sequence is replicated once, homologous recombination takes palce between a copy of the newly replicated inverted sequence and the copy yet to be replicated. This results in a flip of the replicating forks (small arrows) so that they are now chasing each other. Further replication results in generation of amplified arrays of inverted duplications derived from a single initiation of DNA replication during a single S-phase.

accompanying episome formation, as in the first step of *Figure 9*. Evidence for such a deletion is difficult to obtain in most systems since cell lines and tumor cells are often hyperdiploid and deletion in one chromosome can be masked by the presence of additional copies of the locus in the remaining homologous chromosomes. However, it was possible to assess whether deletion occurred by using cell lines containing transfected genes initially present in only one chromosome and in Chinese hamster cells with only a single copy of the endogenous DHFR gene. When episome-containing cells from such cell lines were passaged without selection, clones completely devoid of transfected sequences (176; J.C.Ruiz and G.M.Wahl, in preparation) or the endogenous DHFR gene (B.E.Windle, B.Draper, and G.M.Wahl, in preparation) were observed. Since karyotypic and molecular analyses failed to reveal chromosome loss, deletion coincident with episome formation is favored in these cases.

It remains to be determined how often episome formation is mediated by deletion as opposed to mechanisms which leave the chromosomal locus intact. Deletions equivalent in size to the episomes detected thus far have been proposed to arise by recombination across the bases of the looped replication domains thought to exist in mammalian cells (242,243) and circular molecules containing a replication origin could easily be one product of such a recombination event. Formation of episomes by deletion can account for the formation of inverted novel joints, found in two well-studied cases (196,244). As episomes are acentromeric, amplification can result simply from their unequal segregation at mitosis. Coordinate deletion and episomal formation is attractive because only one or two recombination events are required to account for extrachromosomal amplification. This model may also be relevant to chromosomal amplification. As mentioned previously, DMs can be derived from episomes in some cases. If episome oligomerization to DMs is extensive in a few cells of a population, reintegration of sequences that were extrachromosomal initially could lead to formation of ECRs. Direct evidence for integration of episomes (or DMs) has been presented recently by Carroll *et al.* (176).

10.2.2 The sister chromatid exchange model

In a recent cytological study of the expansion of an ECR, unequal sister chromatid exchanges were observed within the ECR and it was concluded that this mechanism was likely to be involved in expansion of an already amplified array (245). In a study of first-step mutants involving CAD gene amplification in Syrian hamster cells, careful attention was paid to isolate independent clonal populations of mutant cells grown for a minimal time after the primary event (99). Many first-step mutants, which had only 3–6 extra copies of the CAD gene, were analyzed and were found to have amplified as much as 10 000 kb of DNA together with each copy of the selected gene (214). Three novel joints were identified in a set of 33

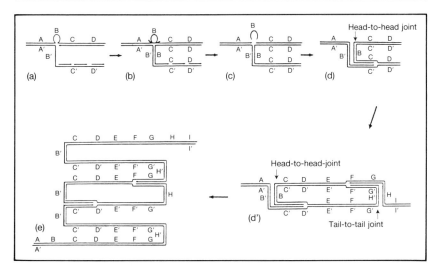

Figure 10. A model for inverted duplication formation and intrachromosomal amplification (reproduced from ref. 224). (a – d) Formation of a head-to-head (C'/BC) joint by copy choice recombination. (d') Structure obtained if a head-to-head (C'/BC) joint and a tail-to-tail (G/H'G') joint are formed by the same events occurring at both ends of the replication bubble. (e) Amplification of the resulting inverted duplication by intrachromosomal double rolling circle replication.

independent mutants and all were present in a single copy per cell. In this case, the large size of each amplified domain would be predicted by the onionskin model only if unscheduled replication occurred over a region containing about 100 replication origins, or would require formation of very large episomes if extrachromosomal DNA were involved. Some of the models described above predict that novel joints can be created and amplified in a single step. Such models may not be relevant in this case since no amplified joints have yet been observed. In particular, both of the episome models predict that the novel joints will be amplified as much as the selected gene. Unequal sister chromatid exchange, already known to be the likely cause of magnification of rRNA genes in *Drosophila* (246), may explain intrachromosomal amplification of very large regions of DNA most easily. This model also predicts explicitly the presence of single copy joints in mutants recovered after only two or three successive recombination events (*Figure 11*). Moreover, the sister chromatid exchange model has relatively strict further predictions which can be tested. In its simplest form, it predicts head-to-tail joining of the amplified domains. Therefore, one expects to see cytologically an abnormally long chromosome in which each extra copy of the selected gene is separated from the rest by a long stretch of co-amplified DNA. The other homolog should be normal, with a single copy of the selected gene. Such a situation is suggested by recent observations (K.A.Smith and G.R.Stark, unpublished data). To generate more than one extra copy per cell, more

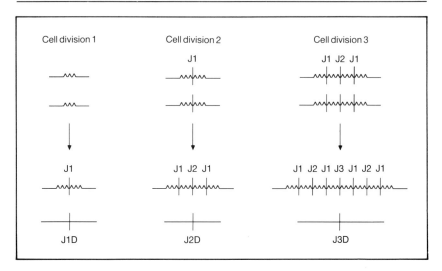

Figure 11. A model for gene amplification in which sister chromatids recombine unequally before mitosis. A lineage is shown in which copy number increases maximally at each cell division. Note that only one pair of sister chromatids is shown and that each daughter cell still has an unaltered chromosome. After the third step there are four copies of J1 and two copies of J2 but only one copy of J3. J2 could also be at single copy number if the increase is not maximal at step three. Thus, this model predicts formation of both single copy and amplified novel joints in head-to-tail arrangements.

than one unequal sister chromatid exchange would have to occur in different cell generations (*Figure 11*). The first event, generating one extra copy, might allow a cell to survive and generate a slow-growing colony and a second event in one cell of this colony could give a new population with more of a selective advantage. Both events would be a part of the operationally defined first step of selection. Examination of very early events should shed more light on these possibilities.

11. Prospects for the future

The advent of new technologies for analyzing large expanses of DNA and new strategies for isolating variants which amplify sequences at high frequency opens up possibilities for identifying the first molecular products of gene amplification and some of the proteins which may participate. Recent work has clearly shown how rapidly the products of amplification can increase in size, rearrange, and integrate into chromosomes. Thus, the focus of future studies should be to develop experimental strategies capable of identifying cells which have amplified a given gene as close to the primary event as possible. Questions concerning the proportion of cases mediated by the initial production of extrachromosomal elements versus those with unstable intrachromosomal expansions can then be answered. Such studies may reveal whether some of the mechanisms presented above can be eliminated from further consideration. If the production of extrachromosomal elements is a preferred mechanism, careful analysis of early events should also reveal how frequently chromosomal deletion generates these elements. Further experiments can then be designed to elucidate whether the sites of recombination that generate deletions are clustered and whether sequences distant from each other in the linear chromosome are brought close together in the higher order structures currently envisioned. Thus, molecular analysis of the products of gene amplification could provide a powerful tool for understanding secondary or tertiary structures of mammalian chromosomes. Such an analysis also has the potential to yield the long sought after origins of mammalian DNA replication.

Powerful new techniques involving two-dimensional gel electrophoresis (65,247,248) should be able to identify a single origin amplified to thousands of copies in a drug-resistant cell line (249,250). Other new methods should be very helpful in the near future. Pulsed-field electrophoresis, which can resolve DNA molecules in the 1 – 10 Mb range, is a powerful tool, and an excellent start has been made in using it to analyze amplified DNA (204). New methods for carrying out *in situ* hybridizations using biotinylated probes with fluorescence detection (251 – 255), which have the ability to resolve single copy sequences on each sister chromatid with high resolution, will be particularly valuable in helping to decide such issues as whether the primary events of first-step amplification take place *in situ* or with movement of an amplified array, how large are the regions involved, and how the evolution of an already amplified array occurs.

The DNA breakage and joining reactions which must be involved in generating amplified DNA clearly require participation of recombination proteins. Does amplification reflect differential expression of such proteins under conditions which induce stress? Alternatively, since DNA replication and gene amplification may be intimately linked, might not the key to understanding lie in the accumulation of DNA damage in metabolically stressed transformed cells since they, unlike normal cells, enter S-phase under suboptimal growth conditions? The high rate of clonal variation in tumor cells due to accumulation of chromosomal alterations could thus derive from an altered capacity for cell-cycle control under conditions of DNA damage, and gene amplification could provide an experimental tool for identifying some of the gene products involved. Isolation of a gene or genes which confer the amplificator phenotype may provide identification of such proteins. If yeast mutants with concurrent hyperamplification of independent loci could be selected, the isolation of the genes responsible would be facilitated greatly. Analysis of how mutations affecting cell-cycle progression affect the generation of gene rearrangements could also be facilitated, since many such mutants are currently available in both fission and budding yeasts.

The phenomenology which dominated the study of gene amplification a decade ago has given way to molecular, genetic, and biochemical approaches capable of sorting out the molecular biology of gene amplification. In the process, the prospects have become brighter for gaining greater insight into the architecture of mammalian chromosomes, the sequences or structures which drive their replication, and the proteins which maintain their integrity or participate in their reorganization.

12. References

1. Cowell, J. K. (1982) Double minutes and homogenously staining regions: gene amplification in mammalian cells. *Annu. Rev. Genet.,* **16,** 21.
2. Schimke, R. T. (1982) *Gene Amplification.* Cold Spring Harbor Laboratory, Cold Spring Harbor, NY.
3. Biedler, J. L., Meyers, M. B., and Spengler, B. A. (1983) Homogenously staining regions and double minute chromosomes, prevalent cytogenetic abnormalities of human neuroblastoma cells. *Adv. Cell. Neurobiol.* **4,** 267.
4. George, D. L. (1984) Amplifications of cellular proto-oncogenes in tumours and tumour cell lines. *Cancer Surveys,* **3,** 497.
5. Hamlin, J. L., Milbrandt, J. D., Heintz, N. H., and Azizkahn, J. D. (1984) DNA sequence amplification in mammalian cells. *Int. Rev. Cytol.,* **90,** 31.
6. Schimke, R. T. (1984) Gene amplification, drug resistance and cancer. *Cancer Res.,* **44,** 1735.
7. Schimke, R. T. (1984) Gene amplification in cultured animal cells. *Cell,* **37,** 705.
8. Stark, G. R. and Wahl, G. M. (1984) Gene amplification. *Annu . Rev. Biochem.,* **53,** 447.
9. Stark, G. R. (1986) DNA amplification in drug resistant cells and in tumours. *Cancer Surveys,* **5,** 1.
10. Gudkov, A. V. and Kopnin, B. P. (1987) Gene amplification and multidrug resistance. *Sov. Sci. Rev. D. Physiochem. Biol.,* **7,** 95 (in English).
11. Schimke, R. T. (1988) Gene amplification in cultured cells. *J. Biol. Chem.,* **263,** 5989.
12. Meuth, M. (1989) Illegitimate recombination in mammalian cells. In *Mobile DNA.* Berg, D. E. and Howe, M. (eds), American Society for Microbiology, Washington DC.
13. Wahl, G. M. (1989) The importance of circular DNA in mammalian gene amplification. *Cancer Res.,* **49,** 1333.
14. Stark, G. R., Debatisse, M., Giulotto, E., and Wahl, G. M. (1989) Recent progress in understanding mechanisms of mammalian DNA amplification. *Cell,* **57,** 901.
15. Blackburn, E. (1982) DNA termini in ciliate macronuclei. *Cold Spring Harbor Symp. Quant. Biol.,* **47,** 1195.
16. Yao, M. -C., Zhu, S. -G., and Yao, C. H. (1985) Gene amplification in *Tetrahymena thermophila*: formation of extrachromosomal palindromic genes coding for rRNA. *Mol. Cell. Biol.,* **5,** 1260.
17. Rochaix, J. -D., Bird, A., and Bakken, A. (1974) Ribosomal RNA gene amplification by rolling circles. *J. Mol. Biol.,* **87,** 473.
18. Wellauer, P. K., Reeder, R. H., Dawid, I. B., and Brown, D. D. (1976) The arrangement of length heterogeneity in repeating units from amplified and chromosomal rDNA from *Xenopus laevis. J. Mol. Biol.,* **105,** 487.
19. Zalokar, M. and Erk, I. (1976) Division and migration of nuceli during early embryogenesis of *Drosophila melanogaster. J. Microbiol. Cell.,* **25,** 97.
20. Foe, V. and Alberts, B. M. (1983) Studies of nuclear and cytoplasmic behaviour during the five mitotic cycles that precede gastrulation in *Drosophila* embryogenesis. *J. Cell Sci.,* **61,** 31.
21. Campos-Ortega, J. and Hartenstein, V. (1987) *The Embryonic Development of Drosophila Melanogaster.* Springer-Verlag, Berlin.
22. Gatti, M., Pimpinelli, S., Bove, C., Baker, B. S., Smith, D., *et al.* (1984) Genetic control

of mitotic cell divison in *Drosophila melanogaster*. In *Proceedings of the 15th International Congress of Genetics*, Vol. 3. New Delhi Oxford/IBH, p. 193.

23. Glover, D. M. (1989) Mitosis in *Drosophila*. *J. Cell. Sci.*, **92**, 137.
24. Freeman, M. and Glover, D. M. (1987) The *gnu* mutation of *Drosophila* causes inappropriate DNA synthesis in unfertilised and fertilised eggs. *Genes Develop.*, **1**, 924.
25. Freeman, M., Nusslein-Volkhard, C., and Glover, D. M. (1986) The dissociation of nuclear and centrosomal division in *gnu*, a mutation causing giant nuclei in *Drosophila*. *Cell*, **46**, 457.
26. Kriegstein, H. J. and Hogness, D. S. (1974) Mechanism of DNA replication of *Drosophila* chromosomes: structure of replication forks and evidence for bidirectionality. *Proc. Natl. Acad. Sci. USA*, **71**, 135.
27. Blumenthal, A., Kreigstein, H., and Hogness, D. (1973) The units of DNA replication in *Drosophila melanogaster* chromosomes. *Cold Spring Harbor Symp. Quant. Biol.*, **38**, 205.
28. McKnight, S. L. and Miller, O. L. Jr. (1977) Electron microscopic analysis of chromatin replication in the cellular blastoderm of *Drosophila melanogaster* embryos. *Cell*, **12**, 795.
29. Steinemann, M. (1981) Chromosomal replication in *Drosophila virilis*. II. Organization of active origins in diploid brain cells. *Chromosoma*, **82**, 267.
30. Hammond, M. P. and Laird, C. D. (1985) Control of DNA replication and spatial distribution of defined DNA sequences in salivary gland cells of *Drosophila melanogaster*. *Chromosoma*, **91**, 267.
31. Hammond, M. P. and Laird, C. D. (1985) Chromosome structure and DNA replication in nurse and follicle cells of *Drosophila melanogaster*. *Chromosoma*, **91**, 279.
32. Dickson, E., Boyd, J. B., and Laird, C. D. (1971) Sequence diversity of polytene chromosome DNA from *Drosophila hydei*. *J. Mol. Biol.*, **61**, 615.
33. Endow, S. A. and Gall, J. (1975) Differential replication of satellite DNA in polyploid tissues of *Drosophila virilis*. *Chromosoma*, **50**, 175.
34. Renkawitz-Pohl, R. and Kunz, W. (1975) Underreplication of satellite DNAs in polyploid ovarian tissue of *Drosophila virilis*. *Chromosoma*, **49**, 375.
35. Nazimiec, M. and Beckingham, K. (1986) 3B55: a repetitious sequence family which is transcribed and proportionally replicated in germ-line polyploid nuclei of *Calliphora erythrocephala*. *Dev. Biol.*, **115**, 398.
36. Beckingham, K. (1982) Insect rDNA. In *The Cell Nucleus*,, vol 10. Busch, H. and Rothblum,L. (eds), Academic Press, New York, p. 206.
37. Spear, B. B. and Gall, J. G. (1973) Independent control of fibrosomal gene replication in polytene chromosomes of *Drosophila melanogaster*. *Proc. Natl. Acad. Sci. USA*, **70**, 1359.
38. Spear, B. B. (1974) The genes for ribosomal RNA in diploid and polytene chromosomes of *Drosophila melanogaster*. *Chromosoma*, **48**, 159.
39. Glover, D. M. (1981) The rDNA of *Drosophila melanogaster*. *Cell*, **26**, 297.
40. Endow, S. A. and Glover, D. M. (1979) Differential replication of ribosomal gene repeats in polytene nuclei of *Drosophila*. *Cell*, **17**, 597.
41. Endow, S. A. (1980) On ribosomal gene compensation in *Drosophila*. *Cell*, **22**, 149.
42. Endow, S. A. (1982) Polytenization of the ribosomal genes on the X and Y chromosomes of *Drosophila melanogaster*. *Genetics*, **100**, 375.
43. Endow, S. A. (1983) Nucleolar dominance in polytene cells of *Drosophila*. *Proc. Natl. Acad. Sci. USA*, **80**, 4427.
44. Beckingham, K. and Thompson, N. (1982) Under-replication of intron[+] rDNA cistrons in polyploid nurse cell nuclei of *Calliphora erythrocephala*. *Chromosoma*, **87**, 177.
45. Spierer, A. and Spierer, P. (1984) Similar level of polyteny in bands and interbands of *Drosophila* giant chromosomes. *Nature*, **307**, 176.
46. Lifschytz, E. (1983) Sequence replication and banding organization in the polytene chromosomes of *Drosophila melanogaster*. *J. Mol. Biol.*, **164**, 17.
47. Breuer, M. E. and Pavan, D. (1955) Behavior of polytene chromosomes of *Rhynchosciara angelae* at different stages of larval development. *Chromosoma*, **7**, 371.
48. Rudkin, G. T. and Collett, S. L. (1957) Disproportionate synthesis of DNA in a polytene chromosome region, *Proc. Natl. Acad. Sci. USA*, **43**, 964.
49. Ficq, A. and Pavan, C. (1957) Autoradiography of polytene chromosomes of *Rhynchosciara angelae* at different stages of larval development. *Nature*, **180**, 983.
50. Gabrusewycz-Garcia,N. (1964) Cytological and autoradiographic studies in *Sciara coprophila* salivary gland chromosomes. *Chromosoma*, **15**, 312.

51. Crouse, H. V. and Keyl, H. G. (1968) Extra replication in the DNA puffs of *Sciara coprophila*. *Chromosoma,* **25**, 357.
52. Glover, D. M., Zaha, A., Stocker, A. J., *et al.* (1982) Gene amplification in *Rhynchosciara* salivary gland chromosomes. *Proc. Natl. Acad. Sci. USA,* **79**, 2947.
53. Millar, S., Hayward, D. C., Read, C. A., *et al.* (1985) Segments of chromosomal DNA from *Rhynchosciara americana* that undergo additional rounds of DNA replication in the salivary gland DNA puffs have only weak ARS activity in yeast. *Gene,* **34**, 81.
54. Lara, F. J. S., Miller, S. E., Hayward, D. C., Read, C. A., Browne, M. J., Santelli, R. V., Garcia-Vallejo, F., Pueyo, M., Zaha, A., and Glover, D. M. (1985) Gene amplification in the DNA puffs of *Rhynchosciara americana* salivary gland chromosomes. In *Cellular Regulation and Malignant Growth*. Ebashi, S. (ed.), Springer-Verlag, Berlin.
55. Spradling, A. and Mahowald, A. P. (1981) A chromosome inversion alters the pattern of specific DNA replication in *Drosophila* follicle cells. *Cell,* **27**, 203.
56. Orr, W., Komitopoulou, K., and Kafatos, F. (1984) Mutants supressing in *trans* chorion gene amplification in *Drosophila*. *Proc. Natl. Acad. Sci. USA,* **81**, 3773.
57. Spralding, A. C. (1981) The organization and amplification of two clusters of *Drosophila* chorion genes. *Cell,* **27**, 193.
58. Osheim, Y. N. and Miller, O. L (1983) Novel amplification and transcriptional activity of chorion genes in *Drosophila melanogaster* follicle cells. *Cell,* **33**, 543.
59. Spralding, A. C., deCicco, D. V., Wakimoto, B. T., Levine, J. F., Kalfayan, L. J., and Cooley, L. (1987) Amplification of the X-linked chorion gene cluster requires a region upstream from the S38 chorion gene. *EMBO J.,* **6**, 1045.
60. Wakimoto, B., Kalfayan, K., and Spradling, A. (1986) Localization of regions controlling *Drosophila* chorion gene expression. *Symp. Soc. Dev. Biol.,* **44**, 43.
61. de Cicco, D. and Spradling, A. (1984) Localization of a *cis*-acting element responsible for the developmentally regulated amplification of *Drosophila* chorion genes. *Cell,* **38**, 45.
62. Delidakis, C. and Kafatos, F. C. (1990) Amplification of a chorion gene cluster in *Drosophila* is subject to multiple *cis*-regulatory elements and to long range position effects. *J. Mol. Biol.,* in press.
63. Orr-Weaver, T. and Spradling, A. C. (1986) *Drosophila* chorion gene amplification requires an upstream region regulating s18 transcription. *Mol. Cell. Biol,* **6**, 4624.
64. Mariani, B. D., Lingappa, J. R., and Kafatos, F. C. (1988) Temporal regulation in development: negative and positive *cis*-regulators dictate the precise timing of expression of a *Drosophila* chorion gene. *Proc. Natl. Acad. Sci. USA,* **85**, 3029.
65. Brewer, B. J. and Fangman, W. L. (1987) The localization of replication origins on ARS plasmids in *S. cerevisiae*. *Cell,* **51**, 463.
66. Snyder, P. B., Galanopoulos, V. K., and Kafatos, F. C. (1986) *Trans*-acting amplification mutants and other eggshell mutants of the third chromosome in *Drosophila melanogaster*. *Proc. Natl. Acad. Sci. USA,* **83**, 3341.
67. Gatti, M., Smith, D. S., and Baker, B. S. (1983) A gene controlling condensation of heterochromatin in *Drosophila melanogaster*. *Science,* **221**, 83.
68. Anderson, R. P. and Roth, J. R. (1977) Tandem genetic duplications in phage and bacteria. *Annu. Rev. Microbiol.,* **31**, 473.
69. Hightower, R. C., Wong, M. L., Ruiz-Perez, L., and Santi, D. V. (1987) Electron microscopy of amplified DNA forms in antifolate-resistant *Leishmania*. *J. Biol. Chem.,* **262**, 14618.
70. Detke, S., Chaudhuri, G., Kink, J. A., and Chang, K. -P. (1988) DNA amplification in tunicamycin-resistant *Leishmania mexicana*. *J. Biol Chem.,* **267**, 3418.
71. Devonshire, A. L. and Sawicki, R. M. (1979) Insecticide-resistant *Myzus persicae* as an example of evolution by gene duplication. *Nature,* **280**, 140.
72. Mouchès, C., Pasteur, N., Berge, J. B., Hyrien, O., Raymond, M., De Saint Vincent, B. R., De Silvestri, M., and Georghiou, G. P. (1986) Amplification of an esterase gene is responsible for insecticide resistance in a California *Culex* mosquito. *Science,* **233**, 778.
73. Field, L. M., Devonshire, A. L., and Forde, B. G. (1988) Molecular evidence that insecticide resistance in peach-potato aphids (*Myzus persicae Sulz.*) results from amplification of an esterase gene. *Biochem. J.,* **251**, 309.
74. ffrench-Constant, R. H., Devonshire, A. L., and White, R. P. (1988) Spontaneous loss and reselection of resistance in extremely resistant *Myzus persicae* (*Sulzer*). *Pest. Biochem. Physio.,* **30**, 1.

75. Prody, C. A., Dreyfus, P., Zamir, R., Zakut, H., and Soreq, H. (1989) *De novo* amplification within a 'silent' human cholinesterase gene in a family subjected to prolonged exposure to organophosphorus insecticides. *Proc. Natl. Acad. Sci. USA,* **86**, 690.

76. Emanuel, J. R., Garetz, S., Schneider, J., Ash, J. F., Benz, E. J. Jr, and Levenson, R. (1986) Amplification of DNA sequences coding for the Na.K-ATPase α-subunit in ouabain-resistant C⁺ cells. *Mol. Cell. Biol.,* **6**, 2476.

77. Wright, J. A., McClarty, G. A., Lewis, W. H., and Srinivasan, P. R. (1988) Hydroxyurea and related compounds. In *Drug Resistance in Mammalian Cells..* Gupta, R. S. (ed.), CRC Press, Inc., Boca Raton, FL.

78. Imam, A. M. A., Crossley, P. H., Jackman, A. L., and Little, P. F. R. (1987) Analysis of thymidylate synthase gene amplification and of mRNA levels in the cell cycle. *J. Biol. Chem.,* **262**, 7368.

79. Collart, F. R., and Huberman, E. (1987) Amplification of the IMP dehydrogenase gene in Chinese hamster cells resistant to mycophenolic acid. *Mol. Cell. Biol.,* **7**, 3328.

80. Debatisse, M., Saito, I., Buttin, G., and Stark, G. R. (1988) Preferential amplification of rearranged sequences near amplified adenylate deaminase genes. *Mol. Cell. Biol.,* **8**, 17.

81. Lewis, A. D., Hickson, I. D., Robson, C. N., Harris, A. L., Hayes, J. D., Griffiths, S. A., Manson, M. M., Hall, A. E., Moss, J. E., and Wolf, C. R. (1988) Amplification and increased expression of alpha class glutathione S-transferase-encoding genes associated with resistance to nitrogen mustards. *Proc. Natl. Acad. Sci. USA,* **85**, 8511.

82. Gerlach, J. H., Endicott, J. A., Juranka, P. F., Henderson, G., Sarangi, F., Deuchars, K. L., and Ling, V. (1986) Homology between P-glycoprotein and a bacterial haemolysin transport protein suggests a model for multidrug resistance. *Nature,* **324**, 485.

83. Gerlach, J. H., Kartner, N., Bell, D. R., and Ling, V. (1986) Multidrug resistance. *Cancer Surveys,* **5**, 25.

84. Roninson, I. G. (1987) Molecular mechanism of multidrug resistance in tumor cells. *Clin. Physiol. Biochem.,* **5**, 140.

85. Moscow, J. A. and Cowan, K. H. (1988) Multidrug resistance. *J. Natl. Cancer Inst.,* **80**, 14.

86. Chabner, B. A. and Gottesman, M. M. (1988) Meeting highlights: William Guy Forbeck Foundation think tank on multidrug resistance in cancer chemotherapy. *J. Natl. Cancer Inst.,* **80**, 391.

87. Gottesman, M. M. and Pastan, I. (1988) The multidrug transporter, a double-edged sword. *J. Biol. Chem.,* **263**, 12163.

88. Riordan, J. R., Deuchars, K., Kartner, N., Alon, N., Trent, J., and Ling, V. (1985) Amplification of P-glycoprotein genes in multidrug-resistant mammalian cell lines. *Nature,* **316**, 817.

89. Gros, P., Croop, J., Roninson, I., Varshavsky, A., and Housman, D. E. (1986) Isolation and characterization of DNA sequences amplified in multidrug-resistant hamster cells. *Proc. Natl. Acad. Sci. USA,* **83**, 337.

90. Roninson, I. G., Chin, J. E., Chjoi, K., Gros, P., Housman, D. E., Fojo, A., Shen, D. -W., Gottesman, M. M., and Pastan, I. (1986) Isolation of human mdr DNA sequences amplified in multidrug-resistant KB carcinoma cells. *Proc. Natl. Acad. Sci. USA,* **83**, 4538.

91. Shen, D. -W., Fojo, A., Chin, J. E., Roninson, I. B., Richert, N., Pastan, I., and Gottesman, M. M. (1986) Human multidrug-resistant cell lines: increased mdr1 expression can precede gene amplification. *Science,* **232**, 643.

92. Lemontt, J. F., Azzaria, M., and Gros, P. (1988) Increased mdr gene expression and decreased drug accumulation in multidrug-resistant human melanoma cells. *Cancer Res.,* **48**, 6348.

93. Jongsma, A. P. M., Spengler, B. A., Van der Bliek, A. M., Borst, P., and Biedler, J. L. (1987) Chromosomal localization of three genes coamplified in the multidrug-resistant CHᴿC5 Chinese hamster ovary cell line. *Cancer Res.,* **47**, 2875.

94. Van der Blieck, A. M., Baas, F., Van der Velde-Koerts, T., Biedler, J. L., Meyers, M. B., Ozols, R. F., Hamilton, T. C., Joenje, H., and Borst, P. (1988) Genes amplified and overexpressed in human multidrug-resistant cell lines. *Cancer Res.,* **48**, 5927.

95. Choi, K., Chen, C., Driegler, M., and Roninson, I. B. (1988) An altered pattern of cross-resistance in multidrug-resistant human cells results from spontaneous mutations in the *mdr*1 (P-glycoprotein) gene. *Cell,* **53**, 519.

96. Scudder, S. A., Brown, J. M., and Sikic, B. I. (1988) DNA cross-linking and cytotoxicity of the alkylating cyanomorpholino derivative of doxorubicin in multidrug-resistant cells. *J. Natl. Cancer Inst.*, **80**, 1294.
97. Kempe, T. D., Swyryd, E. A., Bruist, M., and Stark, G. R. (1976) Stable mutants of mammalian cells that overproduce the first three enzymes of pyrimidine nucleotide biosynthesis. *Cell,* **9**, 541.
98. Johnston, R. N., Beverley, S. M., and Schimke, R. T. (1983) Rapid spontaneous dihydrofolate reductase gene amplification shown by fluorescence-activated cell sorting. *Proc. Natl. Acad. Sci. USA,* **80**, 3711.
99. Zieg, J., Clayton, C. E., Ardeshir, F., Giulotto, E., Swyryd, E. A., and Stark, G. R. (1983) Properties of single-step mutants of Syrian hamster cell lines resistant to *N*-(phosphonacetyl)-L-aspartate. *Mol. Cell. Biol.,* **3**, 2089.
100. Sager, R., Gardi, I. K., Stephens, L., and Grabowy, C. T. (1985) Gene amplification: an example of accelerated evolution in tumorigenic cells. *Proc. Natl. Acad. Sci. USA,* **82**, 7015.
101. Otto, E., McCord, S., and Tlsty, T. D. (1989) Increased incidence of CAD gene amplification in tumorigenic rat lines as an indicator of genomic instability of neoplastic cells. *J. Biol. Chem.,* **264**, 3390.
102. Chambers, A. F., Harris, J. F., and Grundy, J. S. (1988) Rates of generation of methotrexate-resistant variants in cells temperature-sensitive for malignant transformation. *Som. Cell. Mol. Genet.,* **14**, 253.
103. Srivastava, A., Norris, J. S., Shmookler Reis, R. J., and Goldstein, S. (1985) c-Ha-*ras*-1 proto-oncogene amplification and overexpression during the limited replicative life span of normal human fibroblasts. *J. Biol. Chem.,* **260**, 6404.
104. Turner, D. R., Morley, A. A., Haliandros, M., Kutlaca, R., and Sanderson, B. J. (1985) *In vivo* somatic mutations in human lymphocytes frequently result from major gene alterations. *Nature,* **315**, 343.
105. Turner, D. R., Grist, S. A., Janatipour, M., and Morley, A. A. (1988) Mutations in human lymphocytes commonly involve gene duplication and resemble those seen in cancer cells. *Proc. Natl. Acad. Sci. USA,* **85**, 3189.
106. Alitalo, K. and Schwab, M. (1986) Oncogene amplification in tumor cells. *Adv. Cancer Res.,* **47**, 235.
107. Nishimura, S. and Sekiya, T. (1987) Human cancer and cellular oncogenes. *Biochem. J.,* **243**, 313.
108. Masuda, H., Battifora, H., Yokota, J., Meltzer, S., and Cline, M. J. (1987) Specificity of proto-oncogene amplification in human malignant disease. *Mol. Biol. Med.,* **4**, 213.
109. Bonilla, M., Ramirez, M., Lopez-Cueto, J., and Gariglio, P. (1988) *In vivo* amplification and rearrangement of c-*myc* oncogene in human breast tumors. *J. Nat. Cancer Inst.,* **80**, 665.
110. Guerin, M., Barrois, M., Terrier, M. -J., Spielmann, M., and Riou, G. (1988) Overexpression of either c-*myc* or c-*erb*B-2/*neu* proto-oncogenes in human breast carcinomas: correlation with poor prognosis. *Oncogene Res.,* **3**, 21.
111. Johnson, B. E., Makuch, R. W., Simmons, A. D., Gazdar, A. F., Burch, D., and Cashell, A. W. (1988) *myc* family DNA amplification in small cell lung cancer patients' tumors and corresponding cell lines. *Cancer Res.,* **48**, 5163.
112. Yokota, J., Wada, M., Yoshida, T., Noguchi, M., Teraskai, T., Shimosato, Y., Sugimura, T., and Terada, M. (1988) Heterogeneity of lung cancer cells with respect to the amplification and rearrangement of myc family oncogenes. *Oncogene,* **2**, 607.
113. Slamon, D. J., Clark, G. M., Wong, S. G., Levin, W. J., Ullrich, A., and McGuire, W. L. (1987) Human breast cancer: correlation of relapse and survival with amplification of the HER-2/neu oncogene. *Science,* **235**, 177.
114. Van der Vijver, M. J., Mooi, W. J., Wisman, P., Peterse, J. L., and Nusse, R. (1988) Immunohistochemical detection of the neu protein in tissue sections of human breast tumors with amplified neu DNA. *Oncogene,* **2**, 175.
115. Zhou, D., Battifora, H., Yokota, J., Yamamoto, T., and Cline, M. J. (1987) Association of multiple copies of the c-*erb*B-2 oncogene with spread of breast cancer. *Cancer Res.,* **47**, 6123.
116. Slamon, D. J. and Clark, G. M. (1988) Amplification of c-*erb*-2 and aggressive human breast tumors. *Sceince,* **240**, 1795.
117. Yokota, J., Yamamoto, T., Miyajima, N., Toyoshima, K., Nomua, N., Sakamoto, H.,

Yoshida, T., Terada, M., and Sugimura, T. (1988) Genetic alterations of the c-erbB-2 oncogene occur frequently in tubular adenocarcinoma of the stomach and are often accompanied by amplification of the v-erbA homologue. *Oncogene*, **2**, 283.

118. Wong, A. J., Bigner, S. H., Bigner, D. D., Kinzler, K. W., Hamilton, S. R., and Vogelstein, B. (1987) Increased expression of the epidermal growth factor receptor gene is malignant gliomas in invariably associated with gene amplification. *Proc. Natl. Acad. Sci. USA*, **84**, 6899.

119. Yamazaki, H., Fuki, Y., Ueyama, Y., Tamaoki, N., Kawamoto, T., Taniguchi, S., and Shibuya, M. (1988) Amplification of the structurally and functionally altered epidermal growth factor receptor gene (c-erbB) in human brain tumors. *Mol. Cell. Biol.*, **8**, 1816.

120. Humphrey, P. A., Wong, A. J., Vogelstein, B., Freidman, H. S., Werner, M. H., Bigner, D. D., and Bigner, S. H. (1988) Amplification and expression of the epidermal growth factor receptor gene in human glioma xenografts. *Cancer Res.*, **48**, 2231.

121. Malden, L. T., Novak, U., Kaye, A. H., and Burgess, A. W. (1988) Selective amplification of the cytoplasmic domain of the epidermal growth factor receptor gene in glioblastoma multiforme. *Cancer Res.*, **48**, 2711.

122. Ro, J., North, S. M., Gallick, G. E., Hortobagyi, G. N., Gutterman, J. U., and Blick, M. (1988) Amplified and overexpressed epidermal growth factor receptor gene in uncultured primary human breast carcinoma. *Cancer Res.*, **48**, 161.

123. Wang, L., Vass, W., Gao, C., and Chang, K. S. S. (1987) Amplification and enhanced expression of the c-Ki-ras2 protooncogene in human embryonal carcinoma. *Cancer Res.*, **47**, 4192.

124. Adelaide, J., Mattei, M. -G., Marics, I., Raybaud, F., Planche, J., de Lapeyriere, O., and Birnbaum, D. (1988) Chromosomal localization of the *hst* oncogene and its co-amplification with the *int.*2 oncogene in a human melanoma. *Oncogene*, **2**, 413.

125. Zhou,D.J., Casey,G., and Cline,M.J. (1988) Amplification of human int-2 in breast cancers and squamous carcinomas. *Oncogene*, **2**, 279.

126. Lavialle,C., Modjtahedi,N., Cassingena,R., and Brison,O. (1988) High c-*myc* amplification level contributes to the tumorigenic phenotype of the human breast carcinomas cell line SW613-S. *Oncogene*, **3**, 335.

127. Martinsson, T., Stahl, F., Pollwein, P., Wenzel, A., Levan, A., Schwab, M., and Levan, G. (1988) Tumorigenicity of SEWA murine cells correlates with degree of c-*myc* amplification. *Oncogene*, **3**, 437.

128. Ling, V., Chambers, A. F., Harris, J. F., and Hill, R. P. (1985) Quantitative genetic analysis of tumor progression. *Cancer Met. Rev.*, **4**, 173.

129. Nowell, P. C. (1986) Mechanisms of tumour progression. *Cancer Res.*, **46**, 2203.

130. Nicolson, G. L. (1987) Tumour cell instability, diversification and progression to the metastatic phenotype: from oncogene to oncofetal expression. *Cancer Res.*, **47**, 1473.

131. Bishop, J. M. (1987) The molecular genetics of cancer. *Science*, **235**, 305.

132. Temin, H. M. (1988) Evolution of cancer genes as a mutation-driven process. *Cancer Res.*, **48**, 1697.

133. Waghorne, C., Thomas, M., Lagarde, A., Kerbel, R. S., and Breitman, M. L. (1988) Genetic evidence for progressive selection and overgrowth of primary tumors by metastatic cell subpopulations. *Cancer Res.*, **48**, 6109.

134. Cillo, C., Dick, J. E., Ling, V., and Hill, R. P. (1987) Generation of drug-resistant variants in metastatic B16 mouse melanoma cell lines. *Cancer Res.*, **47**, 2604.

135. Damen, J. E., Tagger, A. Y., Greenberg, A. H., and Wright, J. A. (1989) Generation of metastatic variants in populations of mutator and amplificator mutants. *J. Natl. Cancer Inst.*, **81**, 628.

136. Aldaz, C. M., Conti, C. J., O'Connell, J., Yuspa, S. H., Klein-Szanto, A. J. P., and Slaga, T. J. (1986) Cytogenetic evidence for gene amplification in mouse skin carcinogenesis. *Cancer Res.*, **46**, 3565.

137. Gitelman, I., Dexter, D. F., and Roder, J. C. (1987) DNA amplification and metastasis of the human melanoma cell line MeWo. *Cancer Res.*, **47**, 3851.

138. Bevacqua, S. J., Greeff, C. W., and Hendrix, M. J. C. (1988) Cytogenetic evidence of gene amplification as a mechanism for tumor cell invasion. *Som. Cell. Mol. Genet.*, **14**, 83.

139. McIntyre, P. and Stark, G. R. (1988) A quantitative method for analyzing specific DNA sequences directly from whole cells. *Anal. Biochem.*, **174**, 209.

140. Rossman, T. G. and Rubin, L. M. (1988) Colony hybridization to identify mammalian cells containing amplified, transfected or expressed sequences. *Som. Cell. Mol. Genet.,* **14**, 321.

141. Fukumoto, M. and Roninson, I. G. (1986) Detection of amplified sequences in mammalian DNA in-gel renaturation and SINE hybridization. *Som. Cell. Mol. Genet.,* **12**, 611.

142. Roninson, I. G. (1987) Use of in-gel DNA renaturation for detection and cloning of amplified genes. *Methods Enzymol.,* **151**, 332.

143. Fairchild, C. R., Ivy, S. P., Kao-Shan, C. -S., Whang-Peng, J., Rosen, N., Israel, M. A., Melera, P. W., Cowan, K. H., and Goldsmith, M. E. (1987) Isolation of amplified and overexpressed DNA sequences from adriamycin-resistant human breast cancer cells. *Cancer Res.,* **47**, 5141.

144. Fukumoto, M., Shevrin, D. H., and Roninson, I, B. (1988) Analysis of gene amplification in human tumor cell lines. *Proc. Natl. Acad. Sci. USA,* **85**, 6846.

145. Kinzler, K. W., Zehnbauer, B. A., Brodeur, G. M., Seeger, R. C., Trent, J. M., Meltzer, P. S., and Vogelstein, B. (1986) Amplification units containing human N-*myc* and c-*myc* genes. *Proc. Natl Acad. Sci. USA,* **83**, 1031.

146. Cahilly-Snyder, L., Yang-Feng, T., Francke, U., and George, D. L. (1987) Molecular analysis and chromosomal mapping of amplified genes isolated from a transformed mouse 3T3 cell line. *Som. Cell. Mol. Genet.,* **13**, 235.

147. Shiloh, Y., Rose, E., Colletti-Feener, C. Korf, B., Kunkel, L. M., and Latt, S. A. (1987) Rapid cloning of multiple amplified nucleotide sequences from human neuroblastoma cell lines by phenol emulsion competitive DNA reassociation. *Gene,* **51**, 53.

148. Pellegrini, S. and Basilico, C. (1987) Amplification and expression of foreign genes in cells producing polyoma virus large T-antigen. *Oncogene Res.,* **1**, 23.

149. Lacy, J., Summers, W. P., Watson, M., Glazer, P. M., and Summers, W. C. (1987) Amplification and deregulation of MYC following Epstein – Barr virus infection of a human B-cell line. *Proc. Natl. Acad. Sci. USA,* **84**, 5838.

150. Klingel, R., Mincheva, A., Kahn, T., Gissmann, L., Dippold, W., Myer zum Buschenfelde, K. -H., and zur Hausen, H. (1987) An amplification unit in human melanoma cells showing partial homology with sequences of human papillomavirus type 9 and with nuclear antigen 1 of the Epstein – Barr virus. *Cancer Res.,* **47**, 4485.

151. Tokino, T., Fukushige, S., Nakamura, T., Nagaya, T., Murotsu, T., Shiga, K., Aoki, N., and Matsubara, K. (1987) Chromosomal translocation and inverted duplication associated with integrated hepatitis B virus in hepatocellular carcinomas. *J. Virol.,* **61**, 3848.

152. Hatada, I., Tokino, T., Ochiya, T., and Matsubara, K. (1988) Co-amplification of integrated hepatitis B virus DNA and transforming gene *hst*-1 in a hepatocellular carcinoma. *Oncogene,* **3**, 537.

153. Brown, P. C., Tlsty, T. D., and Schimke, R. T. (1983) Enhancement of methotrexate resistance and dihydrofolate reductase gene amplification by treatment of mouse 3T6 cells with hydroxurea. *Mol Cell. Biol.,* **3**, 1097.

154. Hoy, C. A., Rice, G. C., Kovacs, M., and Schimke, R. T. (1987) Over-replication of DNA in S phase Chinese hamster ovary cells after DNA synthesis inhibition. *J. Biol. Chem.,* **262**, 11927.

155. Tlsty, T. D., Brown, P. C., and Schimke, R. T. (1985) UV radiation facilities methotrexate resistance and amplification of the dihydrofolate reductase gene in cultured 3T6 mouse cells. *Mol. Cell. Biol.,* **4**, 1050.

156. Kleinberger, T., Flint, Y. B., Blank, M., Etkin, S., and Lavi, S. (1988) Carcinogen-induced *trans* activation of gene expression. *Mol. Cell. Biol.,* **8**, 1366.

157. Lee, T. -C., Tanaka, N., Lamb, P. W., Gilmer, T. M., and Barrett, J. C. (1988) Induction of gene amplification by arsenic. *Science,* **241**, 79.

158. Rice, G. C., Hoy, C., and Schimke, R. T. (1986) Transeint hypoxia enhances the frequency of dihydrofolate reductase gene amplification in Chinese hamster ovary cells. *Proc. Natl. Acad. Sci. USA,* **83**, 5978.

159. Lücke-Huhle, C. and Herrlich, P. (1987) Alpha-radiation-induced amplification of integrated SV40 sequences is mediated by a *trans*-acting mechanism. *Int. J. Cancer,* **39**, 94.

160. Pasion, S. G., Hartigan, J. A., Kumar, V., and Biswas,D.K. (1987) DNA sequence responsible for the amplification of adjacent genes. *DNA,* **6**, 419.

161. Bantel-Schaal,U. and zur Hausen,H. (1988) Dissociation of carcinogen-induced SV40

DNA-amplification and amplification of AAV DNA in a Chinese hamster cell line. *Virology,* **166**, 113.

162. Yalkinoglu, A. O., Heilbronn, R., Burkle, A., Schleihofer, J. R., and zur Hausen, H. (1988) DNA amplification of adeno-associated virus as a response to cellular genotoxic stress. *Cancer Res.,* **48**, 3123.

163. Hahn, P., Morgan, W. F., and Painter, R. B. (1987) The role of acentric chromosome fragments in gene amplification. *Som. Cell. Mol. Genet.,* **13**, 597.

164. Mariani, B. D. and Schimke, R. T. (1984) Gene amplification in a single cell cycle in Chinese hamster ovary cells. *J. Biol Chem.,* **259**, 1901.

165. Hahn, P., Kapp, L. N., Morgan, W. F., and Painter, R. B. (1986) Chromosomal changes without DNA overproduction in hydroxyurea-treated mammalian cells: implications for gene amplification. *Cancer Res.,* **46**, 4607.

166. Morgan, W. F., Bodycote, J., Fero, M. L., Hahn, P. J., Kapp, L. N., Pantelias, G. E., and Painter, R. B. (1986) A cytogenetic investigation of DNA rereplication after hydroxyurea treatment: implications for gene amplification. *Chromosoma (Berlin),* **93**, 191.

167. Painter, R. B., Young, B. R., and Kapp, L. N. (1987) Absence of DNA overreplication in Chinese hamster cells incubated with inhibitors of DNA synthesis. *Cancer Res.,* **47**, 5595.

168. Creasey, D. C. and Ts'o, P. O. P. (1988) DNA replication in Syrian hamster cells transiently exposed to hydroxyurea. *Cancer Res.,* **48**, 6298.

169. Schimke, R. T., Hoy, C., Rice, G., Sherwood, S. W., and Schumacher, R. I. (1988) Enhancement of gene amplification by perturbation of DNA synthesis in cultured mammalian cells. In *Cancer Cells 6: Eukaryotic DNA Replication.* Cold Spring Harbor Laboratories, Cold Spring Harbor, New York, pp. 317–323.

170. Sherwood, S. W. Schumacher, R. I., and Schimke, R. T. (1988) Effect of cycloheximide on development of methotrexate resistance of Chinese hamster ovary cells treated with inhibitors of DNA synthesis. *Mol. Cell. Biol.,* **8**, 2822.

171. Giulotto, E., Knights, C., and Stark, G. R. (1987) Hamster cells with increased rates of DNA amplification, a new phenotype. *Cell,* **48**, 837.

172. McMillan, T. J., Kalebic, T., Stark, G. R., and Hart, I. R. (1989) High frequency of double drug resistance in a maligant melanoma cell line: implications for combination chemotherapy. Submitted.

173. Rice, G. C., Ling, V., and Schimke, R. T. (1987) Frequencies of independent and simultaneous selection of Chinese hamster cells for methotrexate and doxorubicin (adriamycin) resistance. *Proc. Natl. Acad. Sci. USA,* **84**, 9261.

174. Rolfe, M., Knights, C., and Stark, G. R. (1988) Somatic cell genetic studies of amplificator cell lines. In *Cancer Cells 6: Eukaryotic DNA replication.* Cold Spring Harbor Laboratories, Cold Spring Harbor, New York, p. 325.

175. Schimke, R. T., Sherwood, S. W., Hill, A. B., and Johnston, R. N. (1986) Over-replication and recombination of DNA in higher eukaryotes: potential consequences and biological implications. *Proc. Natl. Acad. Sci. USA,* **83**, 2157.

176. Carroll, S. M., DeRose, M. L., Gaudray, P., Moore, C. M., Needham-Vandevanter, D. R., Von Hoff, D. D., and Wahl, G. M. (1988) Double minute chromosomes can be produced from precursors derived from a chromosomal deletion. *Mol. Cell. Biol.,* **8**, 1525.

177. Ottaggio, L., Agnese, C., Bonatti, S., Cavolina, P., De Ambrosis, A., Degan, P., Di Leonardo, A., Miele, M., Randazzo, R. M., and Abbondandolo, A. (1988) Chromosome aberrations associated with CAD gene amplification in Chinese hamster cultured cells. *Mut. Res.,* **199**, 111.

178. Yandell, D. W., Dryja, T. P., and Little, J. B. (1986) Somatic mutations at a heterozygous autosomal locus in human cells occur more frequently by allele loss than by intragenic structural alterations. *Som. Cell Mol. Genet.,* **12**, 255.

179. Simon, A. E. and Taylor, M. W. (1980) High-frequency mutation at the adenine phosphoribosyltransferase locus in Chinese hamster ovary cells due to deletion of the gene. *Proc. Natl. Acad. Sci. USA,* **80**, 810.

180. Adair, G. M., Stallings, R. L., Nairn, R. S., and Siciliano, M. J. (1983) High-frequency structural gene deletion as the basis for functional hemizygosity of the adenine phosphoribosyltransferase locus in Chinese hamster ovary cells. *Proc. Natl. Acad. Sci. USA,* **80**, 5961.

181. Nalbantoglu, J., Hartley, D., Phear, G., Tear, G., and Meuth, M. (1986) Spontaneous

deletion formation at the aprt locus of hamster cells: the presence of short sequence homologies and dyad symmetries at deletion termini. *EMBO J., 5*, 1199.

182. Nalbantoglu, J. and Meuth, M. (1986) DNA amplification – deletion in a spontaneous mutation of the hamster *aprt* locus: structure and sequence of the novel joint. *Nucleic Acids Res., 14*, 8361.

183. Lehrman, M. A., Schnieder, W. J., Sudhof, T. C., Brown, M. S., Goldstein, J. L., and Russell, D. W. (1985) Mutation in LDL receptor: Alu – Alu recombination deletes exons encoding transmembrane and cytoplasmic domains. *Science, 227*, 140.

184. Lehrman, M. A., Goldstein, J. L., Russell, D. W., and Brown, M. S. (1987) Duplication of seven exons in LDL receptor gene caused by alu – alu recombination in a subject with familial hypercholesterolemia. *Cell, 48*, 827.

185. Cavenee, W. K., Dryja, T. P., Phillips, R. A., Benedict, W. F., Godbout, R., Gallie, B. L., Murphree, A. L., Strong, L. C., and White, R. L. (1983) Expression of recessive alleles by chromosomal mechanisms in retinoblastoma. *Nature, 305*, 779.

186. Koufos, A., Hansen, M. F., Lampkin, B. C., Workman, M. L., Copeland, N. G., Jenkins, N. A., and Cavenee, W. K. (1984) Loss of alleles at loci on human chromosome 11 during genesis of Wilms' tumor. *Nature, 309*, 170.

187. Reeve, A. E., Housiaux, P. J., Gardner, R. J. M., Chewings, W. E., Grindley, R. N., and Millow, L. J. (1984) Loss of a Harvey *ras* allele in sporadic Wilms' tumor. *Nature, 309*, 174.

188. Fearon, E. R., Votgelstein, B., and Feinberg, A. P. (1984) Somatic deletion and duplication of genes on chromsome 11 in Wilms' tumors. *Nature, 309*, 176.

189. Orkin, S. H., Goldman, D. S., and Sallan, S. E. (1984) Development of homozygosity for chromosome 11p markers in Wilms' tumor. *Nature, 309*, 172.

190. Wahl, G. M., Vitto, L., Padgett, R. A., and Stark, G. R. (1982) Single-copy and amplified CAD genes in Syrian hamster chromosomes localized by a highly sensitive method for *in situ* hybridization. *Mol. Cell. Biol., 2*, 308.

191. Levan, G. and Levan, A. (1982) Transition of double minutes into homogeneously staining regions and C-bandless regions in the SEWA tumor. In *Gene Amplification.* Schimke,R.T. (ed.), Cold Spring Harbor Laboratory, Cold Spring Harbor, New York, pp. 91 – 97.

192. Schwab, M., Ramsay, G., Alitalo, K., Varmus, H. E., Bishop, J. M., Martinson, T., Levan,G., and Levan,A. (1985) Amplifications and enhanced expression of the c-*myc* oncogene in mouse SEWA tumor cells. *Nature, 315*, 345.

193. Cherif, D., Lavialle, C., Modjtahedi, N., Le Coniat, M., Berger, R., and Brison, O. (1989) Selection of cells with different chromosomal localizations of the amplified c-*myc* gene during *in vivo* and *in vitro* growth of the breast carcinomas cell line SW613-S. *Chromosoma, 97*, 327.

194. Kaufman, R. J., Brown, P. C., and Schimke, R. T. (1981) Loss and stabilization of amplified dihydrofolate reductase genes in mouse sarcoma S-180 cell lines. *Mol Cell. Biol., 1*, 1084.

195. Meinkoth, J., Killary, A. M., Fournier, R. E. K., and Wahl, G. M. (1987) Unstable and stable CAD gene amplification: importance of flanking sequences and nuclear environment in gene amplification. *Mol. Cell. Biol., 7*, 1415.

196. Ruiz, J. C. and Wahl, G. M. (1988) Formation of an inverted duplication can be an initial step in gene amplification. *Mol. Cell. Biol., 8*, 4302.

197. Biedler, J. L. (1982) Evidence for transient or prolonged extrachromosomal existence of amplified DNA sequences in antifolate-resistant, vincristine-resistant and human neuroblastoma cells. In *Gene Amplification.* Schimke,R.T. (ed.). Cold Spring Harbor Laboratories, Cold Spring Harbor, New York, pp. 39 – 45.

198. Kohl, N. E., Kanda, N., Schrick, R. R., Bruns, G., Latt, S. A., Gilbert, F., and Alt, F. W. (1983) Transposition and amplification of oncogene-related sequences in human nueroblastomas. *Cell, 31*, 359.

199. Thomas, K. R. and Capecchi, M. R. (1987) Site-directed mutagenesis by gene targeting in mouse embryo-derived stem cells. *Cell, 52*, 503.

200. Lin, C. C., Alitalo, K., Schwab, M., George, D., Varmus, H. E., and Bishop, J. M. (1985) Evolution of karyotypic abnormalities and c-*myc* oncogene amplification in human colonic carcinoma cell lines. *Chromosoma, 92*, 11.

201. Trent, J., Meltzer, O., Rosenblum, M., Harsh, G., Kinzler, K., Marshal, R., Feinberg, A., and Volgelstein, B. (1986) Evidence for rearrangement, amplification and expression of c-*myc* in a human glioblastoma. *Proc. Natl. Acad. Sci. USA, 83*, 470.

202. Hamkalo, B. A., Farnham, P. J., Johnston, R., and Schimke, R. T. (1985) Ultrastructural features of minute chromosomes in a methotrexate-resistant mouse 3T3 cell line. *Proc. Natl. Acad. Sci. USA,* **82**, 1126.

203. Borst, P., Van der Bliek, A. M., Van der Velde-Koerts, T., and Hes, E. (1987) Structure of amplified DNA, analyzed by pulsed field gradient gel electrophoresis. *J. Cell. Biochem.,* **34**, 247.

204. Kinzler, K. W., Bigner, S. H., Bigner, D. D., Trent, J. M., Law, M. L., O'Brien, S. J., Wong, A. J., and Vogelstein, B. (1987) Identification of an amplified, highly expressed gene in a human glioma. *Science,* **236**, 70.

205. Garvey, E. P. and Santi, D. V. (1986) Stable amplified DNA in drug-resistant *Leishmania* exists as extrachromosomal circles. *Science,* **233**, 535.

206. Murray, A. W. and Szostak, J. W. (1983) Construction of artificial chromosomes in yeast. *Nature,* **305**, 189.

207. Ardeshir, F., Giulotto, E., Zieg, J., Brison, O., Liao, W. S. L., and Stark, G. R. (1983) Structure of amplified DNA in different Syrian hamster cell lines resistant to *N*-(phosphonacetyl)-L-aspartate. *Mol. Cell. Biol.,* **3**, 2076.

208. Caizzi, R. and Bostock, C. J. (1982) Gene amplification in methotrexate-resistant mouse cells. IV. Different DNA sequences are amplified in different resistant lines. *Nucleic Acids Res.,* **10**, 6597.

209. Fojo, A. T., Whang-Peng, J., Gottesman, M. M., and Pastan, I. (1985) Amplification of DNA sequences in human multidrug-resistant KB carcinoma cells. *Proc. Natl Acad. Sci. USA,* **82**, 7661.

210. Schilling, J., Beverley, C., Simonsen, C., Crouse, G., Setzer, D., Feagin, J., McGrogan, M., Kohlmiller, N., and Schimke, R. T. (1982) Dihydrofolate reductase gene structure and variable structure of amplified DNA structure in mouse cell lines. In *Gene Amplification*. Schimke, R. T. (ed.), Cold Spring Harbor Laboratory, Cold Spring Harbor, New York, pp. 149–153.

211. Shiloh, Y., Shipley, J., Brodeur, G. M., Bruns, G., Korf, B., Donlon, T., Seeger, R., Sakai, K., and Latt, S. A. (1985) Differential amplification, assembly and relocation of multiple DNA sequences in human neuroblastomas and neuroblastoma cell lines. *Proc. Natl. Acad. Sci. USA,* **82**, 3761.

212. Weith, A., Winking, H., Brackmann, B., Boldyreff, B., and Traut, W. (1987) Microclones from a mouse germ line HSR detect amplification and complex rearrangements of DNA sequences. *EMBO J.,* **6**, 1295.

213. Gudkov, A. V., Chernova, O. B., Kazarov, A. R., and Kopnin, B. P. (1987) Cloning and characterization of DNA sequences amplified in multidrug-resistant Djungarian hamster and mouse cells. *Som. Cell. Mol. Genet.,* **13**, 609.

214. Giulotto, E., Saito, I., and Stark, G. R. (1986) Structure of DNA formed in the first step of CAD gene amplification. *EMBO J.,* **15**, 2115.

215. Debatisse, M., Hyrien, O., Petit-Koskas, E., Robert de Saint Vincent, B., and Buttin, G. (1986) Segregation and rearrangement of coamplified genes in different lineages of mutant cells that overproduce adenylate deaminase. *Mol. Cell. Biol.,* **6**, 1776.

216. Saito, I. and Stark, G. R. (1986) Charomids: cosmid vectors for efficient cloning and mapping of large or small restriction fragments. *Proc. Natl. Acad. Sci. USA,* **83**, 8664.

217. Looney, J. E. and Hamlin, J. L. (1987) Isolation of the amplified dihydrofolate reductase domain from methotrexate-resistant Chinese hamster ovary cells. *Mol. Cell. Biol.,* **7**, 569.

218. Ma, C., Looney, J. E., Leu, T. -H., and Hamlin, J. L. (1988) Organization and genesis of dihydrofolate reductase amplicons in the genome of a methotrexate-resistant Chinese hamster ovary cell line. *Mol. Cell. Biol.,* **8**, 2316.

219. Looney, J. E., Ma, C., Leu, T. H., Flintoff, W. F., Troutman, W. B., and Hamilin, J. L. (1988) The dihydrofolate reductase amplicons in different methotrexate-resistant chinese hamster cell lines have at least a 273-kilobase core sequence, but the amplicons in some cell lines are much larger and are remarkably uniform in structure. *Mol. Cell. Biol.,* **8**, 5268.

220. Zehnbauer, B. A., Small, D., Brodeur, G. M., Seeger, R., and Vogelstein, B. (1988) Characterization of N-*myc* amplification units in human neuroblastoma cells. *Mol. Cell. Biol.,* **8**, 522.

221. Shiloh, Y., Korf, B., Kohl, N. E., Sakai, K., Brodeur, G. M., Harris, P., Kanda, N., Seeger, R. C., Alt, F., and Latt, S. A. (1986) Amplification and rearrangement of DNA

sequences from the chromosomal region 2p24 in human neuroblastomas. *Cancer Res.*, **46**, 5297.

222. Biedler, J. L., Chang, T., Scott, K. W., Melera, P. W., and Spengler, B. A. (1988) Chromosomal organization of amplified genes in multidrug-resistant Chinese hamster cells. *Cancer Res.*, **48**, 3179.

223. Ford, M., Davies, B., Griffiths, M., Wilson, J., and Fried, M. (1985) Isolation of a gene enhancer within an amplified inverted duplication after 'expression selection'. *Proc. Natl. Acad. Sci. USA*, **82**, 3370.

224. Hyrien, O., Debatisse, M., Buttin, G., and Robert de Saint Vincent, B. (1988) The multicopy appearance of a large inverted duplication and the sequence at the inversion joint suggest a new model for gene amplification. *EMBO J.*, **7**, 407.

225. Passananti, C., Davies, B., Ford, M., and Fried, M. (1987) Structure of an inverted duplication formed as a first step in a gene amplification event: Implications for a model of gene amplification. *EMBO J.*, **6**, 1697.

226. Ford, M. and Fried, M. (1986) Large inverted duplications are associated with gene amplification. *Cell*, **45**, 425.

227. Saito, I., Groves, R., Giulotto, E., Rolfe, M., and Stark, G. R. (1989) Evolution and stability of chromosomal DNA coamplified with the CAD gene. *Mol. Cell. Biol.*, **9**, 2445.

228. Hyrien, O., Debatisse, M., Buttin, G., and Robert de Saint Vincent, B. (1987) A hotspot for novel amplification joints in a mosaic of Alu-like repeats and palindromic $A + T$ rich DNA. *EMBO J.*, **6**, 2401.

229. Federspiel, N. A., Beverley, S. M., Schilling, J. W., and Schimke, R. T (1984) Novel DNA rearrangements are associated with dihydrofolate reductase gene amplification. *J. Biol. Chem.*, **259**, 9127.

230. Latt, S. A., Lalande, M., Donlon, T., Wyman, A., Rose, E., Shiloh, Y., Korf, B., Muller, U., Sakai, K., Kanda, N., Kang, J., Stroh, H., Harris, P., Bruns, G., Wharton, R., and Kaplan, L. (1986) DNA-based detection of chromosome deletion and amplification: diagnostic and mechanistic significance. *Cold Spring Harbor Symp. Quant. Biol.*, **L1**, 299.

231. Legouy, E., Fossar, N., Llomond, G., and Brison, O. (1989) Structure of four amplified DNA novel joints. *Som. Cell. Mol. Genetics*, **15**, 309.

232. Wahl, G. M., Robert de Saint Vincent, B., and DeRose, M. L. (1984) Effect of chromosomal position on amplification of transfected genes in animal cells. *Nature*, **307**, 516.

233. Carroll, S. M., Gaudray, P., DeRose, M. L., Emery, J. F., Meinkoth, J. L., Nakkim, E., Subler, M., Von Hoff, D. D., and Wahl, G. M. (1987) Characterization of an episome produced in hamster cells that amplify a transfected CAD gene at high frequency: functional evidence for a mammalian replication origin. *Mol. Cell. Biol.*, **7**, 1740.

234. Wahl, G M., Padgett, R. A., and Stark, G. R. (1979) Gene amplification causes overproduction of the first three enzymes of UMP synthesis in N-(phosphonacetyl)-L-aspartate-resistant hamster cells. *J. Biol. Chem.*, **254**, 8679.

235. Bielder, J. L., Melera, P. W., and Spengler, B. A. (1980) Specifically altered metaphase chromosomes in antifolate-resistant Chinese hamster cells that overproduce dihydrofolate reductase. *Cancer Genet. Cytogenet.*, **2**, 47.

236. Jacob, F. (1977) Evolution and tinkering. *Science*, **196**, 1161.

237. Tartof, K. D. (1975) Redundant genes. *Annu. Rev. Genet.*, **9**, 355.

238. Bullock, P. and Botchan, M. (1982) Molecular events in the excision of SV40 DNA from the chromosomes of cultured mammalian cells. In *Gene Amplification*. Schimke, R. T. (ed.) Cold Spring Harbor Laboratory, Cold Spring Harbor, New York, pp. 215–224.

239. Roberts, J. M., Buck, L. B., and Axel, R. (1983) A structure for amplified DNA. *Cell*, **33**, 53.

240. Futcher, A. B. (1986) Copy number amplification of the $2 \mu m$ circle plasmid of *Saccharomyces cerevisiae*. *J. Theor. Biol.*, **119**, 197.

241. Volkert, F. C. and Broach, J. R. (1986) Site-specific recombination promotes plasmid amplifications in yeast. *Cell*, **46**, 541.

242. Vogelstein, B., Pardoll, D. M., and Coffee, D. S. (1980) Supercoiled loops and eukaryotic DNA replication. *Cell*, **22**, 79.

243. Mirkovitch, J., Mirault, M. -E., and Laemmli, U. K. (1984) Organization of the higher-order chromatin loop: specific DNA attachment sites on nuclear scaffold. *Cell*, **39**, 223.

244. Beverley, S. M., Coderre, J. A., Santi, D. V., and Schimke, R. T. (1984) Unstable

DNA amplifications in methotexate-resistant Leishmania consist of extrachromosomal circles which relocalize during stabilization. *Cell*, **38**, 431.

245. Holden, J. J. A., Hough, M. R., Reimer, D. L., and White, B. N. (1987) Evidence for unequal crossing-over as the mechanism for amplification of some homogeneously staining regions. *Cancer Genet. Cytogenet.*, **29**, 139.

246. Endow, S. A. and Atwood, K. C. (1988) Magnification: gene amplification by an inducible system of sister chromatid exchange. *Trends Genet.*, **4**, 348.

247. Huberman, J. A., Spotila, L. D., Nawotka, K. A., El-Assouli, S. M., and Davis, L. R. (1987) The *in vivo* replication origin of the yeast 2 μm plasmid. *Cell*, **51**, 473.

248. Nawotka, J. A. and Huberman, J. A. (1988) Two-dimensional gel electrophoretic method for mapping DNA replicons. *Mol. Cell. Biol.*, **8**, 1408.

249. Heintz, N. H., Milbrandt, J. -D., Theisen, K. S., and Hamlin, J. L. (1983) Cloning of the initiation region of a mammalian chromosomal replicon. *Nature*, **302**, 439.

250. Burhans, W. C., Selegue, J. E., and Heintz, N. H. (1986) Isolation of the origin of replication associated with the amplified Chinese hamster dihydrofolate-reductase domain. *Proc. Natl. Acad. Sci. USA*, **83**, 7790.

251. Pinkel, D., Straume, T., and Gray, J. W. (1986) Cytogenetic analysis using quantitative, high-sensitivity, fluorescence hybridization. *Proc. Natl. Acad. Sci. USA*, **83**, 2934.

252. Pinkel, D., Landegent, J., Collins, C., Fuscoe, J., Segraves, R., Lucas, J., and Gray, J. (1988) Fluorescence *in situ* hybridization with human chromosome-specific libraries: detection of trisomy 21 and translocations of chromosome 4. *Proc. Natl. Acad. Sci. USA*, **85**, 9138.

253. Lawrence, J. B., Villnave, C. A., and Singer, R. H. (1988) Sensitive high-resolution chromatin and chromosome mapping *in situ*: presence and orientation of two closely integrated copies of EBV in a lymphoma line. *Cell*, **52**, 51.

254. Albertson, D. G., Fishpool, R., Sherrington, P., Nacheva, E., and Milstein, C. (1988) Sensitive and high resolution *in situ* hybridization to human chromosomes using biotin labelled probes: assignment of the human thymocyte CD1 antigen genes to chromsome 1. *EMBO J.*, **7**, 2801.

256. Landegent, J. E., Jansen in de Wal, N., Dirks, R. W., Baas, F., and van der Ploeg, M. (1987) Use of whole cosmid cloned genomic sequences for chromosomal localization by non-radioactive *in situ* hybridization. *Hum. Genet.*, **77**, 366.

Index